S0-ADL-400

Innovation in East Asia

Innovation in East Asia

The Challenge to Japan

Michael Hobday

Senior Fellow, Science Policy Research Unit,
University of Sussex, UK

Edward Elgar

© Michael Hobday 1995

All rights reserved. No part of this publication may be reproduced, stored in a retrieval system, or transmitted in any form or by any means, electronic, mechanical, photocopying, recording, or otherwise without the prior permission of the publisher.

Published by
Edward Elgar Publishing Limited
Gower House
Croft Road
Aldershot
Hants GU11 3HR
England

Edward Elgar Publishing Company
Old Post Road
Brookfield
Vermont 05036
USA

British Library Cataloguing in Publication Data
Hobday, Michael
   Innovation in East Asia: The Challenge to
   Japan
   I. Title
   338.95

Library of Congress Cataloguing in Publication Data
Hobday, Michael.
      Innovation in East Asia : the challenge to Japan / Michael Hobday.
         p.   cm.
      Includes bibliographical references.
      1. Technological innovations—Economic aspects—East Asia.
   2. Technological innovations—Economic aspects—Korea (South).
   3. Technological innovations—Economic aspects—Taiwan.
   4. Technological innovations—Economic aspects—Singapore.
   5. Technological innovations—Economic aspects—Hong Kong.
   6. Technological innovations—Economic aspects—Japan.
   7. Electronic industries—East Asia.   I. Title.
   HC460.5.Z9T44   1995
   338'.064'095—dc20   94–47593
                                                                          CIP

ISBN 1 85898 017 8

Printed and bound in Great Britain by
Biddles Ltd, Guildford and King's Lynn

Contents

Contents

Figures

Tables

Acknowledgements

I would like to express my thanks to the many East Asian friends and colleagues who helped me during my research visits to the region, provided me with insights into how South Korea, Taiwan, Singapore and Hong Kong developed so rapidly, and made my field work so enjoyable and informative. They include: Young Rak Choi, Hwan-Suk Kim, Yong-chan Park and Hang Sik Park in South Korea; C.D Tam and Cornelis Ho in Hong Kong; Ban Seng Tan and Roland Seet; and Wen-Jeng Kuo, Samuel Wang and Jacob Y. H. Jou in Taiwan. I am especially indebted to Dong Jin Koh of Samsung in South Korea for his valued friendship and assistance.

Many of my colleagues in SPRU helped me with their customary support, analytical insights, vigorous criticisms, helpful discussions and their time. In no particular order, they include Martin Bell, Chris Freeman, Keith Pavitt, Margaret Sharp, Jose Cassiolato, William Walker, Mark Dodgson, Norman Clark and Diana Hicks. I am especially thankful to Yao Su Hu for his early guidance and constant interest during the research.

I am also indebted to the UK Economic and Social Research Council who funded much of the research for the book over the two year period 1992 and 1993 (ESRC Research Project Reference: ROOO 23 3116). Finally, I owe a large debt to Sylvia Meli and my two children who put up with me while I wrote the book and spent months abroad.

Foreword

Few people doubt today that a fundamental change is taking place in the balance of world economic power. For two centuries the North Atlantic area was the centre of the world economy and the countries of Western Europe and North America dominated world production, world commerce, world technology and investment flows. In the second half of the twentieth century they are being rapidly overtaken not just by Japan, but by the whole region of Eastern Asia. Many books have been written about Japan and an increasing number about South Korea and the other 'dragons'. The World Bank has published a report entitled 'The East Asian Miracle'. Napoleon's famous remark that China was a sleeping giant which would move the world on awakening has proven prophetic.

However, despite this general recognition of an historic turning point many features of this transition remain unexplained. Michael Hobday's book is an exceptionally valuable contribution to our understanding because of several highly original features. In the first place, he criticizes the 'flying geese' theory and shows that in some respects the success of the four dragons, of China and other Asian countries is a challenge to Japan. No doubt the Asian countries have learnt a great deal from Japan and, as Michael Hobday shows, Japan is the main source of their imports of capital goods and technology. Moreover, the extraordinary growth of the Japanese economy after the Second World War provided an example and an impetus to the entire region. Nevertheless, the very success of their catching-up effort has meant that Japanese firms, handicapped by the high Yen as they have been, now face increasingly intense competition from other East Asian firms throughout the region.

The emphasis is on *firms* because the second major original contribution which Michael Hobday makes is to show *how* firms were able to upgrade their technology and their marketing to compete in increasingly sophisticated products and services. He emphasizes the variety of strategies and management organization and the specific features of each of the Asian dragons.

In particular, his treatment of *Chinese* entrepreneurship and the role of the overseas Chinese communities is an essential complement to the diffusion of Japanese management technologies, which have been the focus of so much attention in this region and indeed worldwide. Michael Hobday's analysis of

xiii

the interdependence between export marketing and the upgrading of technology is an outstanding contribution to the whole theory of 'catching up' in the world economy.

Finally, this book is particularly valuable in showing the key role of the electronics industry in the progress of Japan and most of the other Asian countries. Whilst he denies that any 'leapfrogging' has taken place he does show that all the dragons and now increasingly China and Malaysia have moved much more rapidly than European countries to develop their telecommunication infrastructure and related services and to expand their manufacture of a widening variety of electronic consumer and capital goods. Starting from a very small base and a very low level of industrialization in the 1950s the dragons are now among the leading countries in the world in production and export of electronic goods and services.

There are many other features of this book which make it a uniquely valuable contribution not just to economics but to the way the world is changing at the end of the 20th century.

Christopher Freeman
Science Policy Research Unit
University of Sussex, Brighton

Acronyms

ASEAN	Association of South East Asian Nations
ASIC	application-specific integrated circuit
CAD	computer-aided design
CAM	computer-aided manufacture
CEPED	Council for Economic Planning and Development (Taiwan)
CMOS	complimentary metal oxide semiconductor
CRT	cathode ray tube
DRAM	dynamic random access memory
D-VDR	digital-video disk recorder
EC	European Community
EDB	Economic Development Board (Singapore)
EEPROM	Electrically erasable programmable read only memory
EFTA	European Free Trade Association
ERD	Engineering Research and Development Department (Anam, South Korea)
ERSO	Electronic Research Services Organization
ETRI	Electronics and Telecommunications Research Institute (South Korea)
FDI	foreign direct investment
GFI	government-funded institute
GNP	gross national product
HDD	hard disk drive
HKPC	Hong Kong Productivity Council
HP	Hewlett-Packard
ID	Industry Department (Hong Kong)
III	Institute for Information Industry (Taiwan)
IME	Institute for Micrelectronics (Singapore)
ITRI	Industrial Technology Research Institute (Taiwan)
JIT	just-in-time (delivery)
KFTA	Korean Foreign Trade Association
KIST	Korean Institute for Science and Technology
KMT	Kuomintang (Chinese Nationalist Party)
KOTRA	Korean Trade Promotion Corporation
KTA	Korea Telecom Authority

KTC	Korea Telecom Company
KTRI	Korean Telecommunications Research Institute
LCD	liquid crystal display
MOEA	Ministry of Economic Affairs (Taiwan)
MOS	metal oxide semiconductor
MOST	Ministry of Science and Technology (South Korea)
MPU	microprocessor unit
NIE	newly industrializing economy
NMOS	N-channel metal oxide semiconductor
OBM	own-brand manufacture
ODM	own-design manufacture
OECD	Organization for Economic Cooperation and Development
OEM	original equipment manufacture
OPC	Oriental Precision Company
PABX	private automatic branch exchange
PC	personal computer
P-DIP	plastic dual-in-line (chip) package
PLC	product life cycle
R&D	research and development
RISC	reduced instruction-set chip
ROK	Republic of Korea
SAIT	Samsung's Advanced Institute of Technology
S&T	Science and Technology
SCI	Science Citation Index
SDF	Skills Development Fund (Singapore)
SME	small and medium sized enterprises
SRAM	static random access memory
STIC	Singapore Technology Industrial Corporation
TEAMA	Taiwan Electronic Appliances Manufacturers Association
TDC	Trade Development Council (Hong Kong)
TDS	Time division switching
TDX	Time division exchange
TI	Texas Instruments
TNC	transnational corporation
TSMC	Taiwanese Semiconductor Manufacturing Corporation
UMC	United Microelectronics Corporation
VCR	videocassette recorder
VTC	Vocational Training Council (Hong Kong)

1. Introduction: East Asia's technological development

1.1 THE CHALLENGE TO JAPAN

This book concerns the competitive challenge to Japan arising from within East Asia. Led by firms from the four newly industrializing economies (NIEs) or 'dragons' (South Korea, Taiwan, Hong Kong and Singapore), the challenge involves strategic alliances between Pacific Asian companies and American and European transnational corporations (TNCs) and the spread of manufacturing capability into China and other parts of the region.[1] Following the lead of Japan, the four dragons have forged their way into high-technology markets and become internationally respected suppliers of semiconductors, computers, disk drives and many other products. Other followers in the region, notably Thailand, Malaysia, Indonesia and China, have grown extremely rapidly, largely as a result of export-led industrialization.

Although still in its embryonic form, the challenge to Japan has taken root and is gathering pace. Despite Japan's continuing economic dominance of the region, other Pacific Asian firms have learned to compete in manufacturing exports and generated the capacity to innovate. The technological advance of the four NIEs, in particular, has irrevocably altered the balance of competitive power in the region and opened up new strategic opportunities for TNCs from other parts of the world. Increasingly, the competitive threat to Japan derives from within Pacific Asia, rather than from Europe or the US.

As the book shows, Japan is still overwhelmingly dominant in terms of economic and technological power, outward investment, trade and corporate influence. However, Japan can neither take its traditional economic dominance of East Asia for granted nor can it assume that its manufacturing leadership is unchallenged. In an increasing number of industrial areas, local East Asian companies have gained control over modern manufacturing processes, started to design new products and created distinctive new organization forms for competing on the international stage. The book shows that, although nascent, the challenge is based on a solid historical foundation of learning. Firms and governments within the region have devised an impressive variety of new routes to industrial development. The unprecedented success of these strategies promises to defy Japan in the future and to acceler-

1

ate the spread of manufacturing capacity to China and other fast-growing economies of the region.

The purpose of this book is to provide a measure of the strength of the challenge to Japan, by examining its historical roots and analysing the nature and direction of non-Japanese innovation in the region. The study focuses on the four leading Pacific Asian NIEs. In a number of key industrial areas, including computers and semiconductors, the dragons are increasingly pressurizing Japan from below. The four countries also provide a window onto other dimensions of the competitive dynamics of the region, including Taiwanese sub-contracting alliances with US TNCs, the spread of Hong Kong's manufacturing base into China, Singapore's role as a TNC investment hub for South East Asia and the way in which large South Korean firms are forging strategic partnerships around the globe, partly to overcome their traditional reliance on Japan.

Much of this recent development, and especially export growth, has taken place in electronics. In the electronics industry, firms from the four NIEs learned the technology necessary to overcome market entry barriers and compete on the international stage. Some companies caught up with the international market leaders from Japan and the West. Many local firms acquired and improved upon foreign technologies in advanced consumer goods, printed circuit board assembly, calculators, fax machines and telecommunications products.

In comparison with Japan, there is remarkably little evidence on how firms in the NIEs overcame technological barriers to entry and built bridges into international markets. Most existing studies of the Asian miracle have focused on the debate over the relative importance of market mechanisms *versus* government policies. Some studies stress the importance of state policies in guiding industrialization (Wade 1990; Amsden 1989; Kim and Dahlman 1992; Appelbaum and Henderson 1992; Chowdhury and Islam 1993). Others stress the role of market forces, interest rate policies and macroeconomic stability, claiming that government intervention has been overstressed and cannot explain East Asia's success (World Bank 1993).

The dominance of the market *versus* state debate has left little room for the study of the technology strategies of East Asian firms. The central competition question of how companies acquired and 'learned' technology is almost untouched in the literature on East Asian industrialization.[2] This is a serious oversight as firms are the locus of competition, exports, wealth creation, innovation and productivity growth in East Asia as elsewhere. Although most serious analysts accept the importance of technological acquisition to industrial growth, the paths, patterns and mechanisms of technological development in the NIEs remains a mystery to most observers.

In some respects, the achievements of the four dragons are more remarkable than those of Japan. While the latter experienced a long period of industrialization prior to the Second World War, the four dragons and the second-tier NIEs are genuine latecomers. Thirty years ago each dragon suffered from poverty, unemployment and a backward technological infrastructure. As recently as the early-1960s, the GDP per capitas of South Korea and Taiwan were equivalent to those of the poorer African nations.

Since that time, as a result of rapid export-led industrialization, the GDPs of the four dragons have grown at average rates of between 8 per cent to 10 per cent per annum. This has produced full employment, rising wages and a large increase in East Asian trade. Although the NIEs do not yet pose a competitive threat to Japan in most areas, measured in terms of outward investment, they have already surpassed Japan within East Asia. By acquiring technology and investing in research and development (R&D), local firms have produced ever higher quality goods, forcing Japan into yet more complex, technology-intensive products, especially in electronics. Yet the question of how East Asian firms achieved this competitive edge is still largely unanswered.

1.2   OBJECTIVES OF THE STUDY

The purpose of this book is to explore how firms from the four leading East Asian NIEs acquired technology and managed their entry into international markets. The study identifies the main international channels of technology transfer and shows how the market entry strategies of leading firms have evolved over time. Taking each country in turn, the book analyses the technological learning mechanisms of large and small electronics exporters and asks whether firms innovate or are merely imitators of foreign market leaders.[3] The mechanisms by which companies brought about the diffusion of technology from the advanced countries and set about improving and adapting foreign technology are explored. While it was not possible to examine the second-tier Asian NIEs, some of the synergies between the four dragons and Japan, China and the other growth centres of the region are touched upon.

Although the book focuses mainly on electronics, other fast-growing export industries such as footwear, sewing machines and bicycles are included. As the largest and fastest-growing industry in the region the choice of electronics needs little justification. For the past two decades this industry has been a major contributor to employment, productivity growth, exports and overall industrial development in each of the four dragons. It is also a technologically complex, highly competitive industry where, according to conventional wisdom, developing countries should have little chance of succeeding.

The main theoretical contribution is to introduce and develop the idea of the latecomer firm and to contrast the structures and strategies of developing-country firms with those of advanced-country leaders and followers. The notion of the latecomer firms is used to illustrate how and why local companies forged relationships with foreign manufacturers and buyers in their search for technology and access to markets. As a guiding analytical concept, the latecomer firm is used to explain the origins, structures, strengths and continuing weaknesses of East Asian companies. The research illustrates how these firms fused export marketing and technology transfer channels to overcome their disadvantages on the international stage. By developing an institutional system of export-led technology development, the latecomer firms were able to assimilate, adapt and improve foreign technology.

In addressing the challenge to Japan, the book assesses the scale and depth of innovative capability among the latecomer firms and the changing nature of their alliances with foreign TNCs. During the 1980s, companies from Taiwan and Hong Kong exploited the opening up of China by direct investment and joint ventures. This has added to the competitive potential of the latecomers and contributed directly to the rapid economic growth in China.

Many authors have drawn attention to the importance of government policies to Pacific Asian growth. This book goes further by comparing policy mechanisms, strategies and outcomes across the four countries. Highlighting the plurality of government policies and development models within the region, the study shows that there was no single path to development, nor any single model or lesson for other developing countries. However, there are important principles which led to success and which may be of interest to other developing countries. These are identified and highlighted in the study.

As a result of their success, latecomer firms now face an increasingly complex, sometimes hostile, international technology and market environment. As more East Asian latecomers approach the technology frontier in electronics they have been forced to devise new strategies to overcome their remaining disadvantages to become fully-fledged market competitors. Strategic partnerships, conducted on an equal footing with Western and Japanese leaders, are one means of accessing highly advanced electronics and information technologies. Another is the stepping up of in-house investments in R&D to bring new product innovations to the market. In looking towards the future, the book outlines the prospects and problems of latecomer firms as they face the innovation frontier in electronics.

1.3   RESEARCH METHODS

Given the complexity of the East Asian phenomenon, the research combines elements of economics, innovation theory and management studies, including historical analyses, industrial surveys and case studies. This approach helped explore the dynamics of firms' strategies and show how companies caught up in electronics, grasping the breadth and depth of their achievements. During the study, large amounts of data on economic performance, industrial structure, trade, foreign direct investment (FDI) and technological achievements were analysed to place the company case studies in context. The overall Pacific Asian development picture is summarized in Chapter 2 and each of the single country chapters contains important evidence on government policy, industrial structure and historical achievements.

The core of the field research was a large number of interviews with members of local firms, carried out by the author in each of the four countries between 1990 and 1993. This involved several field trips to the region and more than 130 interviews. Companies were interviewed using a number of open-ended discussion guides to gather information on the origin of the company, key points of structural transformation, mechanisms of technological learning, chief sources of technology and methods for overcoming barriers to entry into international markets. In each country chapter, firm case studies are presented within the historical context of industrial development and government policies.

In addition to latecomer firms, some TNCs and joint ventures were also studied to compare their development and partly to show the links between local and foreign firms. In Singapore the study is largely about TNCs as they were the centrepiece of development in electronics, while in Hong Kong and Taiwan TNCs and joint ventures continued to be important throughout the 1980s and 1990s. In South Korea local companies became the single dominant force for development. The major government-funded technology institutes in each country were visited and analysed[4] as were many of the relevant state agencies and academic groups concerned with science and technology (S&T).

Country and company comparisons are a central part of the research method. By examining differences and similarities between the four dragons, the study was able to draw out basic principles underlying East Asian success in electronics (Chapter 8). The four countries highlighted extreme variety in corporate strategy and orientation, government policies and the role of TNCs. South Korea, for instance, adopted a process-intensive, high-volume approach, heavily dependent upon Japanese companies for electronics technology. By contrast, Taiwan's small specialist firms operated in large numbers of fast-changing niche markets, relying on both American and Japanese companies for technology and key components.

Although the research was restricted mainly to electronics, there was enough variety to make useful comparisons of government policies, industrial structures, corporate ownership, firm technology strategies and mechanisms of learning. This intra-sectoral variety allows the research to contribute to a wider understanding of innovation in Asia Pacific. However, there are limits to the electronics case which are discussed in the conclusion.

1.4   THEORETICAL CONCERNS

The study was influenced by a number of theories and propositions about the nature of technological learning, innovation and diffusion in developing countries. These ideas, alongside the issue of the latecomer firm, are used to guide the study and make sense of the findings. Following previous research, the central argument is that firm-level learning is the chief mechanism by which foreign technology is diffused across national boundaries, between and within firms (Hobday 1990). For sustained foreign technology transfer to occur, various learning mechanisms and channels need to function effectively. There also needs to be a conducive industrial, economic and government policy environment, as discussed in Chapter 8.

One important conceptual issue concerns the nature and direction of innovation in East Asia. Although the linear model of S&T has been widely criticized, conventional 'Western' models of innovation still tend to stress the importance of R&D to innovation. According to received wisdom, R&D is conducted by firms, products are then developed, refined and marketed, according to a product life cycle which flows from early to mature stages and from product-intensive to process-intensive competition (Kotler 1976; Utterback and Abernathy 1975; Utterback and Suarez 1993). These powerful ideas run very deep in the innovation and policy analysis literature. In this work, the timing, sequence and nature of firms' technological activities are analysed in order to contrast the East Asian technological experience with mainstream models of innovation. As Chapter 8 shows, the patterns of the East Asian latecomer innovation were more or less the reverse of the conventional Western models. A new model is therefore introduced to explain patterns witnessed in Pacific Asia.

Another interesting theoretical question is whether or not the process of East Asian innovation concurs with the popular idea of technological leapfrogging. This argues that some developing countries may be able to bypass earlier vintages of technology and enter electronics and information technology during the early phase of the new paradigm (Soete 1985). To address this issue in detail, Chapter 6 uses Singapore as a test case of leapfrogging. Among the dragons, Singapore appears to have benefited most from its information infrastructure and, like the other NIEs, had considerable success in electronics. Singapore is an interesting test case, as the pre-conditions with respect to education and absorptive capacity appeared to be in place for leapfrogging to occur. If it did not occur here, then it is not likely to occur elsewhere. The nature and sequence of electronics development is studied in detail to assess the leapfrogging hypothesis. The chapter shows that the notion of industrial leapfrogging is fundamentally flawed and misleading in its implications for educational and industrial policy for other developing countries.

1.5 BOOK STRUCTURE

The next chapter provides the regional context and economic background for East Asia and the four dragons. It analyses their achievements and shows the dominance of Japan in the region. By examining the strategies and nature of the challenge posed by the four NIEs, a critique of the Japanese-oriented 'flying geese' model of East Asian development is provided. Although popular in Japan, the flying geese model underplays the distinctive business and technological characters of the dragons and their strong outward investment in the second-tier NIEs of the region. Accepting the significance of Japan as a driving force for the region's development, the chapter also stresses the importance of the US both as the largest market for East Asian manufactured goods and as an important additional source of technology for the region.

Chapter 3 examines the literature on technological development in East Asia and puts forward a preliminary model for guiding the empirical study and interpreting the results. Several studies on development in the region are combined to highlight interesting questions concerning the likely phasing, nature, strategies and patterns of technological learning in latecomer firms. In particular, a seminal study of export-marketing by Wortzel and Wortzel (1981) is combined with the technology studies to develop a set of *a priori* arguments as to how technological learning links to export market development in the NIEs. The preliminary model is used as a benchmark for assessing the empirical data.

Chapters 4 to 7 analyse each of the four NIEs, tracing the origins and histories of the electronics industry and pointing to the ways in which governments intervened to support industry. The achievements, strengths and weaknesses of the industries are analysed using data gathered from each country on growth, industrial development, structural change and exports. A sample of individual technological learning strategies of firms is presented for each country, looking at internal learning efforts and chief foreign suppliers of technology. The evidence of latecomer learning and innovation is assessed and continuing weaknesses are pinpointed.

The country chapters show how latecomer companies succeeded in coupling together technology sources and market entry channels, creating the essential demand-pull of international markets and forcing a pattern of continuous improvement on local firms. Arrangements such as joint ventures and OEM (original equipment manufacture) were exploited both as systems for acquiring technology *and* as mechanisms for overcoming barriers into export markets. Local and foreign traders were important technology conduits, as were the subsidiaries of TNCs, through their sub-contracting arrangements and training of local engineers and technicians.

The evidence also shows how latecomer companies institutionalized foreign market-technology channels to enable them to learn the art of manufacturing and become skilled technology developers. Today, these institutional structures are deeper, broader and more robust than ever before. They enable exports to dictate the pace and pattern of technological advance and provide firms with a powerful focusing device for innovation, foreign marketing and industrial investment. In particular, OEM is both an organizational innovation of great importance and an evolving system of market-technology progress for East Asia's electronics industry.

The Conclusion (Chapter 8) distils the key success factors for each country by comparing and contrasting policy mechanisms, industrial structures, firm strategies, corporate ownership and company size. A simple model is presented to characterize the diversity of policy frameworks found in each of the countries and to integrate the empirical findings of the book. The framework emphasizes the plurality of non-Japanese development approaches in East Asia and shows how different governments' policies were employed to achieve similar objectives.

Chapter 8 also addresses the question of how East Asian innovation relates to the conventional models of innovation noted earlier. The preliminary model put forward in Chapter 3 is extended to show how typical latecomer firms succeeded in innovating from behind the technology frontier. In contrast with the normal Western models of R&D-led innovation, latecomer firms typically began with simple assembly tasks and only much later introduced R&D as a significant factor in their innovation efforts. Sub-contracting and OEM acted as a training school for many of the latecomers, enabling them to sell their goods under the brand names of leaders and to acquire skills and know-how.[5] Historically, this is an important new departure for developing countries, allowing masses of latecomers to sub-contract their way up the technology ladder, develop new industries, build innovative capabilities and catch up with the industrialized countries.

Concluding the main themes of the book and drawing lessons for theory and policy, Chapter 8 goes beyond the market *versus* state debate, arguing that the activities and strategies of firms are at the heart of the East Asian development miracle. The skills and abilities of firms cannot be taken for granted by policy makers. These have to be developed, nurtured and encouraged within a stimulating policy environment. In the absence of a sufficient number of skilled entrepreneurs to lead industrialization and technology development, the most carefully planned policy is bound to fail.

In assessing the challenge to Japan the Conclusion looks to the future, showing how latecomer company strategies will need to evolve further to confront a more complex and challenging technology environment. To continue on their rapid growth trajectories, firms will have to make further innovative efforts in research, product design, key components and capital-goods technologies. Increasingly, foreign suppliers of technology will see them as rivals, rather than sub-contractors and OEM suppliers. In response, the latecomers are likely to continue their outward location of low-end activities into China and other countries to further reduce costs and maintain their desirability as finished goods suppliers to foreign TNCs. At the same time, latecomers will seek to increase the share of output sold under their own brand names, thereby capturing more of the value added. Further large programmes of R&D will be needed to generate new product designs, and more innovation partnerships with Western and Japanese firms are likely to take shape in the future. At the centre of the newly emerging challenge to Japan is a mass of creative and fast-growing latecomer companies. As more of the latecomers face the innovation frontier and the markets of China and the other countries of the region expand, the challenge to Japan is likely to grow in stature and effectiveness.

1.6 LESSONS FOR OTHER COUNTRIES

The technological experience of the four dragons will be of interest to the second-tier economies of Pacific Asia including Thailand, Indonesia, Malaysia and China, as they attempt to internalize a technologically dynamic, fast-track industrialization path similar to the four leading NIEs. As East Asia is increasingly recognized as a powerful new world centre of industrial competitiveness and manufacturing innovation, policy makers and analysts in Europe, the US and Japan should take note of how latecomer firms caught up in electronics.

Although there are no simple lessons for other economies, the book may be of interest to developing countries such as Brazil, India, Venezuela and Mexico which are currently attempting to reverse long-term policies of import-substitution and, in some respects, to emulate the success of the dragons in specific areas. While each county has a different history, geography and set of economic opportunities and problems, some of the basic principles regarding the technological advance of East Asia's latecomers may be of interest and value. The book is written in such a way as to be accessible to students, teachers and policy makers concerned with modern economic and technological development.

NOTES

1. The terms East Asia and Pacific Asia are used interchangeably in the book. The geography of the region is reviewed in Chapter 2.
2. The most revealing technology studies tend to examine the general processes of technological change and learning, rather than firm-level strategies (Lall 1982 and 1992; Dahlman et al. 1985; Westphal et al. 1985; Fransman and King 1984). As yet, very few studies compare company strategies across countries to identify similarities and differences. Single country studies, such as Amsden (1989) on South Korea and Wade (1990), Schive (1990) and Dahlman and Sananikone (1990) on Taiwan, tend to focus on overall industrial processes and the role of government, rather than firm strategies. Studies which analyse government policies and historical, economic and business developments tend to overlook technology and the firm (e.g. Vogel 1991; Abegglen 1994; Kelly and London 1989, Haggard 1991; Kwan 1994). Whitley (1992) takes a cultural and sociological approach to understanding business practices in East Asia. A few studies such as Kim and Kim (1989) and Chung and Lee (1989) examine business management issues in South Korea. Jansson (1994) looks at TNCs in South East Asia. The most incisive book yet written on Japanese corporate strategies is by Abbeglen and Stalk (1985). Social and political issues are also dealt with (e.g. Bello and Rosenfeld 1991; Galenson 1992; Dixon and Drakakis-Smith 1993). With few exceptions electronics, the largest sector in the region, is largely ignored in economic assessments (e.g. World Bank 1993; Riedel 1988). One study looks at electronics in the NIEs from an international 'top down' perspective, rather than from the point of view of local firms (Ernst and O'Connor 1992). Wellenius et al. (1993) provide a collection of international policy and industry surveys examining electronics. Some of these studies are touched on during the book.
3. Chapter 3 provides the latecomer analytical framework and defines terms such as innovation, imitation and learning.
4. It is outside the scope of this book to analyse the role of S&T institutes in East Asia. However, the author's evaluation of the largest S&T institutes in each of the four dragons is presented in Rush et al. (1994).
5. The idea of OEM as a harsh industrial training school for Taiwanese companies is put forward by Cowley (1991 p. 20).

2. East Asian regional dynamics

2.1 UNDERSTANDING THE REGION'S PROGRESS

The economic development of the Pacific Asian region is undoubtedly one of the most important events in the world economy in the post-war period. For the first time in history, a group of once-developing countries look set to catch up with the industrialized economies. In the lead, the four dragons have grown rapidly, raised their per capita incomes and improved their technological infrastructures. Other poorer East Asian economies have benefited directly from the achievements of Japan and the four NIEs. Taking the region as a whole, if forecasts are correct, by the year 2000 the Pacific Asian region's GNP (including Japan) will exceed that of the European Union and equal that of North America.

The purpose of this chapter is to illustrate the economic and technological inter-dependence of the region as a whole as essential background for understanding technological advance in the four NIEs. Each dragon is linked into a dynamic regional system of innovation. Studies which ignore the inter-country capital, technological and trade linkages miss one of the crucial underpinnings of the East Asian miracle.[1]

In examining the economic achievements of the region, this chapter focuses on the Japanese-oriented flying geese model of East Asian development. This is the first model to account systematically for inter-country trade and technology connections, attributing much of the growth of the four NIEs and neighbouring East Asian economies, including Malaysia, Thailand, Indonesia and China, to Japan. Although Japan is undeniably important, the chapter argues that the model fails to give due recognition to the distinctive indigenous efforts of the four dragons to accumulate technology and to export overseas. Another difficulty with the model is that it ignores the significance of the US economy, both as a vital export market and as a source of technology for the region.

The chapter shows that by the early 1990s, the four dragons were already an important source of East Asian development, matching Japan's outward investment in the less developed economies of the region. Following divergent development paths, the dragons developed their own distinctive innovation systems. With the partial exception of South Korea, these models differ

11

markedly from the Japanese system of innovation described by many writers.[2]

2.2   ECONOMIC GROWTH

Figure 2.1 shows the economies of Pacific Asia and their populations in 1991. The main industrial centres were Japan, the four dragons, the second-tier NIEs (Thailand, Indonesia, Malaysia and the Philippines) and China.[3] Within East Asia, the South East Coastal region of China grew rapidly, receiving large amounts of DFI from Taiwan and Hong Kong, which contributed substantially to China's rapid economic development (see Section 2.11 below).

Many studies illustrate the extraordinary economic growth of East Asia. The World Bank shows that during the 1970s and 1980s the GNP growth of Japan and the four dragons increased at two to three times the rates of most older industrialized countries. According to *Fortune*, by the year 2000 the GNP of the Asia Pacific Rim will exceed that of the European Union (excluding the EFTA countries) and equal North America's income. Some observers forecast that by 2020, Pacific Asian GDP will double that of the European Union (IEEE 1991 p. 24).

Table 2.1 shows the historical growth of the region, compared with the developed countries, since the 1960s.[4] Hong Kong's GNP grew at more than 8 per cent annually during the period 1960 to 1988. Likewise, Singapore's growth outstripped that of Japan and other developed countries (albeit from a lower base). South Korea and Taiwan grew generally above 8 per cent throughout the 1960s, 1970s and 1980s. By contrast, US growth averaged around 3 per cent, similar to (then) European Economic Community growth (Wade 1990 p. 34). Japanese growth slowed to around 5 per cent during the 1970s and 1980s, while the four dragons continued to average 8 to 9 per cent each year.

One remarkable feature was the growth achieved during and after the oil crises and world recession of the early 1970s. While other countries, including Japan, decelerated, three of the four NIEs increased their pace of growth during the 1970s. Despite the slowdown in most OECD countries in the early 1990s, rapid rates of growth continued among most East Asian economies. Malaysia, Thailand and China also quickened their pace during the recession.

According to the World Bank, East Asia became one of the world's principal engines of growth in 1991, growing by 6.8 per cent, compared with the G-7 average of 1.9 per cent, largely due to expanding intra-regional trade and investment (*Far Eastern Economic Review* 24 September 1992 pp. 89–90). Around 33 per cent of East Asian exports were destined for other countries of

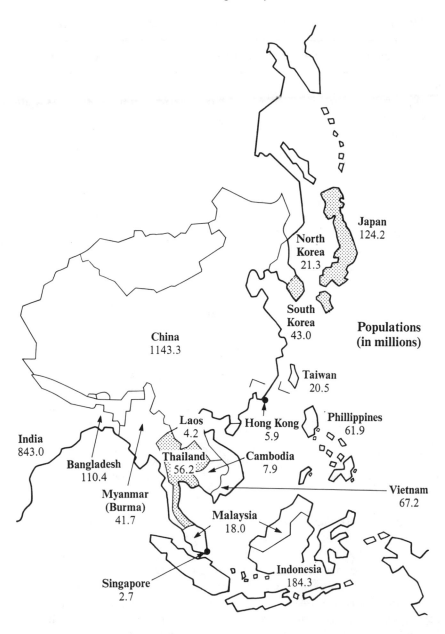

China
1143.3

North
Korea
21.3

Japan
124.2

South
Korea
43.0

**Populations
(in millions)**

Taiwan
20.5

Laos
4.2

Hong Kong
5.9

Phillippines
61.9

India
843.0

Bangladesh
110.4

Thailand
56.2

Cambodia
7.9

Vietnam
67.2

Myanmar
(Burma)
41.7

Malaysia
18.0

Singapore
2.7

Indonesia
184.3

Source: IEEE (1991 p. 25).

Figure 2.1   Profile of the Pacific Asian region

Table 2.1   Average annual rates of growth of real GNP (selected years)

Country/group	1960–69	1970–79	1980–88	1988	1991	1992	1993[a]
Four NIEs							
Hong Kong	10.0	9.4	8.0	10.5	3.9	5.2	5.6
South Korea	7.7	9.5	8.7	15.9	8.4	5.5	6.3
Singapore	8.9	9.5	7.0	10.0	6.7	5.6	6.5
Taiwan	9.5	10.2	7.5	9.5	7.0	6.7	n/a
ASEAN-4							
Indonesia	3.4	7.8	5.8	7.5	7.0	5.5	6.0
Malaysia	6.5	8.1	5.3	6.6	8.6	8.5	8.0
Philippines	3.0	6.3	1.6	6.4	0.0	3.3	1.5
Thailand	8.3	7.4	5.6	7.1	7.9	7.5	7.8
Other Asia							
China	2.9	7.5	9.2	10.2	4.6	12.0	10.0
India	3.7	3.2	5.6	4.4	3.7	4.0	4.8
OECD (selected)							
Canada	5.7	4.7	3.1	4.3	–0.2	0.3	2.8
Japan	10.9	5.2	5.3	5.1	3.1	0.9	2.4
United States	4.1	2.8	2.6	3.6	0.4	2.2	2.5

Note: [a] 1992 official estimates.

Sources: Data for all economies 1960–88 (James 1990 p. 4); Asian economic data for 1991, 1992 and 1993 (*Far Eastern Economic Review* 19 November 1992 pp. 76–7; 14 January 1993 pp. 56–7). OECD data for 1991 to 1993 (*Economist* 21 March 1992 p. 145; 9 May 1992 p. 147; 26 January 1993 p. 139).

the region in 1990 (*Far Eastern Economic Review* 27 August 1992 pp. 54–5). In 1991 intra-regional trade increased by a further 17 per cent.

As James (1990 p. 3) points out, the long-run performance of the four dragons can be explained by high levels of productivity growth, high rates of investment and improved export competitiveness. After the oil shocks, the NIEs (and the ASEAN-4) increased their share of exports to the US and Japan, while exports from Latin America declined. Exports to the Middle East oil-exporting economies also played a part in the NIEs' growth.

As a consequence of fast growth, the four dragons jumped up the GNP per capita rankings. As Table 2.2 shows, the most dramatic improvements were made by Taiwan and South Korea. In 1962 these countries were at the level

Table 2.2 GNP per capita levels and country rankings (US$, current prices) 1962 and 1986

	1962 Rank[a]/ amount	Comparable[b] economies	1986 Rank/ amount	Comparable economies
Taiwan	85/ 170	Zaire Congo, PR	38/ 3,580	Greece Malta
South Korea	99/ 110	Sudan Mauritania	44/ 2,372	Surinam Argentina
Hong Kong	40/ 450	Spain Malta	28/ 6,906	Saudi Arabia Israel
Singapore	38/ 490	Greece Spain	25/ 7,411	New Zealand Bahamas

Notes:
[a] Ranking out of 129 countries.
[b] Comparable' refers to countries immediately above and below the four dragons in the Atlas listing.

Source: Derived from Wade (1990 p. 35) (original data from World Bank Atlas, except for Taiwan).

of African nations such as Zaire and the Sudan. By 1986 the two dragons had moved up the ranking by 47 and 55 places respectively, overtaking other developing countries such as Mexico and Brazil (Wade 1990 p. 34). Although Hong Kong and Singapore began at relatively high per capita levels, they too raced ahead by 12 places and 13 places respectively, overtaking Brazil, Mexico, El Salvador and South Africa, among others.

2.3 REGIONAL DEVELOPMENT

Table 2.3 presents a range of selected indicators to illustrate the development of the four dragons. In 1990, Japan's GNP was overwhelmingly dominant, being 12 times larger than its nearest rival, South Korea, and nearly six times larger than the four dragons combined. South Korea's GNP and population were significantly larger than Taiwan's, although the latter's income per capita was slightly higher than South Korea's.

Table 2.3   Japan and the four dragons: selected economic indicators 1990[a].

	Japan	South Korea	Taiwan	Hong Kong	Singapore
GNP (US$ billion, 1990)	$2,961bn	$238bn	$162bn	$70bn	$35bn
Per Capita GDP[b] (US$000s, 1991)	$13.6k	$5.7k	$6.5k	$14.4k	$13.2k[c]
Population (millions, 1990)	123.5m	42.8m	20.2m	5.8m	2.7m
Trade Balance (US$ billion)	$63.6bn	−$2.0bn	$12.5bn	−$0.34bn	−$5.1bn
Major Export Trade Partner	US	US	US	US	US
Major Import Trade Partner	US	Japan	Japan	China	Japan
Share of Manuf. in GDP (1988)[d]	29%	32%	38%	22%	30%
Inflation Rate (1990)	3.1%	8.6%	4.6%	9.7%	3.4%
People per Auto (1990)	4	28	13	27	10
People per PC (1990)	15	36	20	30	18
People per Telephone (1990)	1.8	3.3	3.0	2.1	2.3
Literacy	100%	92.7%	91.2%	88.1%	82.9%
Stock Market Capitalisation[e] (US$ 1991)	$2,906bn	$116bn	$131bn	$113bn	$52.7bn

Notes:
US billions (a thousand million) are used throughout.
[a]   Unless otherwise stated.
[b]   1988 by purchasing power.
[c]   1991 data.
[d]   Figures cited in Fukasaku (1991 Table 2).
[e]   As of mid-1991.

Sources: *Fortune* (1991 pp. 128–9); except: people per telephone and literacy IEEE (1991 p. 25); share of manufacturing in GDP Fukasaku (1991 Table 2); Singapore per capita income (EDB 1992a p. 12)

Hong Kong and Singapore are city states with small populations. By 1990 their per capita GNPs were on a par with Japan's. Conversely, the GNP per capita of South Korea and Taiwan lagged behind substantially. Inflation rates were high in 1990 in South Korea and Taiwan, due to economic over-heating.[5]

As a general indicator of commitment to education, the high rates of literacy testify to each country's efforts to ensure a well-educated workforce. The supply of engineering graduates is particularly high in each of the drag-ons. Riedel (1988 p. 22) shows the steady improvement in the four dragons' educational attainment levels since 1962.

People per personal computer (PC) and telephone are rough indicators of the level of diffusion of high technology. As Table 2.3 shows, Singapore and Taiwan had high rates of PC diffusion in 1990, compared with South Korea and Hong Kong. South Korea and Taiwan lagged behind the city states in telephone penetration. Singapore was slightly behind Hong Kong in tel-ephones per person, but both economies had very advanced public telecom-munications infrastructures.

As shown below in more detail, the US was the largest export market for each of the countries, including Japan, during the past two decades. NIE exports led to large trade surpluses with the US. By contrast, the dragons have tended to suffer from heavy trade deficits with Japan, due to imports of finished goods, machinery, materials and components. Historically, the US economy sustained the region's export-led growth. Despite the recent in-crease in intra-regional trade, the US still accounted for 24 per cent of total East Asian exports in 1990, down from 30 per cent in 1984 (*Far Eastern Economic Review* 27 August 1992 pp. 54–5).

2.4   TRADE AND TECHNOLOGY

Fukasaku (1991) illustrates the technological advance of the dragons during the 1980s by documenting the shift away from exports of natural resources and labour-intensive goods, to exports of human capital and technology-intensive goods. South Korea's exports of the latter increased from 43 per cent in 1979 to 57 per cent in 1988. Similar progress occurred in each of the four NIEs.

Using data for two years, Fukasaku (1991) shows that despite South Korea's greater size, its total exports were less than Taiwan's in 1979 and 1988. Taiwan's export success is dramatic given its heavy reliance on small and medium sized firms. During the 1980s the tiny overseas Chinese firms matched South Korea's larger, better known *chaebol* in overall export performance. Exports from Singapore and Hong Kong were roughly half that of South

Korea, again an impressive performance given their tiny populations and relatively small GNPs. Strong export achievements contributed to their relatively high GNP per capitas, noted earlier.

James (1990 p. 9) shows that Pacific Asian intra-regional trade grew ninefold between 1970 and 1988, far exceeding the rate of expansion of global trade. By 1988 at least half of the exports of each economy were destined for other Pacific Asian countries. Fukasaku (1991) confirms that East Asia became the most dynamic trading region in the world economy, noting that growth was largely due to the surge in intra-regional trade between Japan and the dragons. During the period 1986 to 1989 the currencies of Taiwan and South Korea appreciated under policy pressure from the US. This stimulated imports by the dragons (mainly from Japan).

By the late 1980s, four NIEs were no longer low-wage economies. Costs had risen substantially and, as incomes rose, large sectors of labour-intensive production were relocated to the second-tier NIEs and China.

2.5   THE IMPORTANCE OF ELECTRONICS

During the 1980s, electronics became the largest export sector in East Asia. The four dragons succeeded in exporting huge volumes of PCs, disk drives, semiconductors, colour TVs and video cassette recorders. In addition, as the country chapters show, each of the dragons developed significant competitive and technological capabilities in at least some areas of advanced electronics.

Table 2.4 shows the increasing share of electronics in total exports from the four NIEs during the latter part of the 1980s.[6] The data show that electronics constituted a rapidly increasing share of manufactured exports from each of the NIEs. By 1988, the dragons together exported around US$52.5 billion worth of electronics, more than the US and Germany, and nearly 73 per cent of Japan's electronics exports. Each NIE exported more electronics than France and Italy and each was catching up with the UK, but still well behind Germany. The data also show the growth of Malaysian electronics exports and the importance of electronics to manufacturing in each of the second-tier NIEs (with the exception of Indonesia). Although South Korea is often seen as the leading NIE electronics exporter, it is interesting to note that each of the other dragons exported equivalent amounts, with smaller populations and GDPs.

Data gathered from each of the dragons (see Chapters 4 to 7) show that the dragons started building their electronics industries in the 1960s and that they took off during the 1980s and into the 1990s. In Hong Kong, exports of electronics increased nearly threefold from US$2.8 billion in 1981 to around US$7.5 billion in 1990. South Korean electronics exports grew from just

Table 2.4   *Electronic exports 1985 and 1988 (US$ billions): selected countries*

	1985	%[a]	1988	%
Four dragons				
South Korea	3.9	15.2	13.6	24.1
Taiwan	4.6	15.9	12.0	21.5
Singapore	4.8	41.4	13.2	45.0
Hong Kong	5.8	36.5	13.7	23.9
Second-tier NIEs				
Malaysia	2.1	50.2	4.6	48.9
Indonesia	0.1	4.5	0.2	3.0
Thailand	0.4	14.3	1.4	16.8
Philippines	0.8	35.2	1.3	30.1
Developed countries				
US	27.8	19.6	45.7	18.6
Japan	42.0	27.0	72.1	27.8
Germany	14.0	9.4	25.6	8.8
UK	10.0	15.6	17.9	15.0
France	6.9	10.4	11.7	9.5
Italy	4.3	6.7	7.2	6.3

Note: [a] As percentage of total manufactured exports.

Sources: Derived from Ernst and O'Connor (1992 pp. 98–9); original data from: *Yearbook of World Electronics Data* (1988 and 1990, Benn/Elsevier Volumes 1 and 2); *Handbook of International Trade and Development Statistics* (1988), United Nations and other sources.

US$2 billion in 1980 to US$20 billion in 1991, overtaking textiles and garments, steel, shipping and automobiles to become the largest single export sector. With a population of just over 3 million, Singapore exported around US$15 billion in 1991, nearly as much as South Korea and more than Taiwan which exported US$12.2 billion.[7]

By the early 1990s, local firms from each NIE were able to design their own electronics products, often working in partnership with buyers and foreign technology suppliers. Local firms upgraded their electronics capabilities and narrowed the technological gap between themselves and the foreign market leaders. Some, such as Samsung Electronics of South Korea and ACER of Taiwan, overtook the traditional market leaders in key areas such as dynamic random access memory (DRAM), semiconductors and computer peripherals. Many firms progressed from simple assembly in the 1960s to more sophisticated activities requiring software and precision engineering skills in the 1980s.

Since the late 1980s, electronics has constituted the largest export sector in each of the four countries, except for Hong Kong where it was the second largest after clothing and textiles. Electronics accounted for about 40 per cent of Singapore's total manufacturing output in 1992, and around 40 per cent of its exports (*Electronics* 26 April 1993 p. 5). In South Korea electronics accounted for around 28 per cent of total exports in 1991.[8] Singapore became the world's largest manufacturer of hard disk drives and a competent producer of complex professional and consumer electronics. South Korea's leading firm, Samsung, achieved world class technological capabilities in semiconductor memories as well as colour TVs, camcorders and compact disk players. Taiwan was the world's largest producer of printed circuit boards for PCs and a major supplier of colour monitors, finished PCs, fax machines and calculators. Hong Kong had significant export successes in fax machines, cordless telephones and workstations.

2.6   THE FLYING GEESE MODEL

Analysing Pacific Asia as an integrated economic region began with the idea of the flying geese, put forward by Prof. A. Kaname in the 1930s and developed by Prof. K. Akamatsu in 1956.[9] This model gives prominence to Japan as the driving economic and technological force in the region. Japan developed first, gaining a strong technological base, then as wages and other costs rose in Japan, production facilities were relocated to the NIEs and to other lower-cost economies.

This process accelerated as Japan's exchange rate appreciated after the so-called Plaza Agreement of the (then) G-5 members in September 1985. Following this agreement, the value of the Yen doubled against the US dollar in two years. The effect was to make Japanese exports more expensive abroad and to encourage outward investment by Japan (Urata 1990 p. 1). After 1985, like an inverted 'V', the economies of Pacific Asia began to fly together towards economic development, led by Japan.

After 1985 there was indeed a massive surge of outward FDI from Japan. In the three year period 1986 to 1988, Japanese investments exceeded the cumulative value of FDI during the previous three and a half decades measured from 1951 to 1985 (Urata 1990 p. 6). Most FDI flowed to North America and other developed countries, but a significant share went to East Asia.

The flying geese model puts Japan at the front of the four dragons. As Japanese wages increased and the Yen appreciated, production facilities and technology flowed outwards from Japan, first to the four NIEs, then to the second-tier ASEAN economies (principally Thailand, Malaysia and Indonesia) and to China. Later, as wage costs and technological levels rose in the dragons, their currencies appreciated and they too increased their outward investment into the second-tier NIEs and China.

According to Yamashita (1991 pp. 2–3) the four dragons owe their export achievements largely to Japanese subsidiaries operating from within their economies and/or joint ventures with Japanese companies. Japan, so the argument goes, led the economic development of the region through trade, aid and FDI. The four NIEs imported not only machines, parts and materials but also management styles from Japan. This view of the past leads to a similar view of the future:

> when we observe the whole region, the East and South-East Asian nations are going to be integrated under the Japanese umbrella, economically and technologically. (Yamashita 1991 p. 4)

A narrower, trade-based version of the flying geese idea is put forward by Fukasaku (1991), following Rana (1990). This version does not deal directly with FDI, technology acquisition or management styles. The development of Pacific Asia is modelled as a trade-led, catching up process. Japan is the leader followed by the other economies at various stages of industrialization. As the more advanced economies progress towards exports of technology-intensive products, they leave room for imports of relatively unskilled, labour-intensive standard products (Fukasaku 1991 pp. 14–16).

Although the various flying geese models attempt to see East Asia as an integrated region, one immediate difficulty is that there is no mention of the overseas Chinese in the region's development.

2.7 OVERSEAS CHINESE MODELS

Most business and economic studies of East Asia focus on the Japanese economy, management and organizational styles. However, cultural, historical and sociological studies of the overseas Chinese (e.g. Mackie 1992; Redding 1991; Whitley 1992) highlight their significance to the economic development of the region.

The populations of Taiwan and Hong Kong are largely made up of ethnic Chinese and Singapore's population is dominated by the Chinese. Taiwan and Hong Kong have developed largely as a result of Chinese business systems, connections and investment capital. Singapore has been run by overseas Chinese but has depended on US, Japanese and European investments.

In contrast with Japan's corporate *keiretsu* ('societies of businesses'), which contain some of the world's largest firms, Taiwan's economy, like Hong Kong's, depends to a far greater degree on thousands of tiny family-owned businesses. Most of these firms operate in a Chinese style, characterized by autocratic patriarchal management, fast response to changing market niches and overseas family connections. Successful companies have grown fairly large but still remain very small compared with Japanese corporations.

Whereas South Korea owes much of its conglomerate structure to the nearby Japanese model, management styles in Taiwan owe as much, or more, to American connections as to Japanese firms.[10] Many of Taiwan's leading industrialists worked in American high technology firms and attended US universities. Often, Taiwanese companies retain strong technological links with US corporations in Silicon Valley and elsewhere.

The more traditional overseas Chinese businesses rely heavily on personal connections sometimes called *guanxi*. As Cowley (1991 p. 8) points out, connections are strongest within the family, then clan, village and Chinese home province. Today's worldwide *guanxi* enable Chinese businesses to match consumer demand in the US with production in China and Taiwan. Typically, the patriarch controls the finances from Hong Kong or Taiwan. His 'number-one-son' manages the factory in China, Thailand or Malaysia. Having been to a university in the US, 'number-two-son' may well work in California, assessing new computer innovations. A large proportion of Hong Kong's outward investment was directed to the Guangdong province of China in search of cheap labour as costs rose in Hong Kong. Around 80 per cent of Hong Kong's Chinese population had relatives in Guangdong.

Given the spread and diversity of overseas Chinese, accurate data on their activities are limited. One report claims that there are around 55 million overseas Chinese, of which 51 million reside in Asia. The 'GNP' of Asia's overseas Chinese in 1990 was estimated at around US$450 billion, around 25 per cent larger than China's economy at that time (*Economist* 18 July 1992 pp. 21–4). Overseas Chinese saving and investment rates were estimated to be extremely high. Savings approximated 25 per cent to 45 per cent of GNP per annum. In 1990, liquid assets (excluding securities) were estimated at around US$1.5 trillion to US$2 trillion, which compared with approximately US$3 trillion for Japan in 1990 (measured in terms of bank deposits) with double the population.

Even if such reports are exaggerated there is little doubt that, in addition to Japan, the overseas Chinese are a powerful force for development in parts of East Asia. In South East Asia, the economies of Malaysia, Indonesia and Thailand each rely on Chinese minorities for a significant share of their economic activity. Indonesia's 7.2 million ethnic Chinese compare with Thailand's 5.8 million and Malaysia's 5.2 million. One study claims that the local Chinese, who account for 4 per cent of the Indonesian population, controlled

17 of the largest 25 business groups (*Economist* 18 July 1992 p. 21). Malaysia attracted around US$2.3 billion in manufacturing investment from Taiwan in 1990, compared with US$3.1 billion from Japan. According to *Time* (14 September 1992 p. 24) cumulative Taiwanese investment in Malaysia totalled US$5 billion in 1992, more than Japan's total investment.[11]

China's recent growth was partly fuelled by large investments from Taiwan and Hong Kong. The economic links between Taiwan, Hong Kong and the South East Coastal region of China (mainly Fu Chien and Guangdong provinces) grew much closer during the 1980s. Hundreds of small Taiwanese firms accept export orders from abroad, manufacture in mainland China and carry out the final shipping from Hong Kong (Gee 1991 p. 34). Total trade (including smuggling) with China through Hong Kong and Taiwan was estimated at around US$10 billion in 1992 (*Financial Times* 13 November 1992 p. 15, and 7 October 1992 p. 6). By 1991 Taiwan had invested a cumulative US$3.4 billion in Mainland China (*China Post* 31 July 1992 p. 9).[12]

Given the significant differences between the development strategies of the three 'Chinese' dragons (Taiwan, Hong Kong and Singapore), it is more accurate to speak of East Asian 'models' of development, rather than one single model. This diversity of development patterns is overlooked in the flying geese model which ignores the role of the overseas Chinese. Another factor overlooked in the model is the importance of the US economy.

2.8   THE ROLE OF THE US ECONOMY

Data on the dragons' major trading partners during the 1980s illustrate the importance of the US economy in pulling the region forward through its thirst for imports.[13] During the 1980s, each of the NIEs enjoyed large trade surpluses with the US. By contrast, each had large bilateral trade deficits with Japan in most years. In fact, Japan imported less from the dragons than the (then) European Community.

Among the four NIEs, Taiwan was the largest exporter to the US, followed closely by South Korea, Hong Kong and Singapore. Under pressure to reduce their surpluses with the US, the dragons diversified successfully into Europe during the 1980s. Exports to Japan also increased, but to a lesser extent.

During this period the four NIEs imported large amounts of technology, machinery and components from Japan, exporting finished goods to the US and Europe. As a result, the four countries' foreign trade balances have for most years been positive, but not very large. Taiwan's balance was around US$10 billion per annum during the latter part of the 1980s. South Korea's balance slipped into deficit in 1985 and 1990, but was positive for most of the decade (Chaponniere 1992 pp. 75–6).

Consistent with the flying geese model, Japan's role has been as a supplier of technology and components to the dragons. However, the US and European Community played a more important part than Japan as importers of finished products during the 1980s.

2.9   TECHNOLOGICAL RELIANCE ON JAPAN

Table 2.5 confirms that the four dragons have emerged as major export markets for Japan. In 1990, East and South East Asia imported around US$92 billion from Japan, compared with US$90 billions for the US. The four NIEs accounted for 62 per cent of East and South East Asian imports from Japan in 1990. South Korea's imports from Japan matched Germany's. Each of the NIEs imported more from Japan than did Britain.

Table 2.5   Exports from Japan: selected countries 1990 (US$ billions)

Country	Imports
US	90.3
Germany	17.8
Britain	10.8
Canada	6.7
S.Korea	17.5
Taiwan	15.4
Hong Kong	13.1
Singapore	10.7
Total NIEs	56.7
Other SE Asia	35.3
Total East Asia	92.0
Total Others	69.3
Total Exports	286.9

Source: *Fortune* (October 7 1991 p. 158).

Leading South Korean and Taiwanese firms found it difficult to break free of dependence on Japanese capital goods and key components, especially in electronics. In 1991 South Korea imported around US$21 billion worth of high-technology industrial goods from Japan, a large proportion of the country's total import bill. This produced a deficit with Japan of US$8.8

billion. Similarly Taiwan's deficit with Japan was around US$9.7 billion in 1991, mostly in machinery and components (Holden and Nakarmi 1992 pp. 24–5).

In response to trade deficits with Japan, the governments of Taiwan and South Korea practised import restriction. In 1991 South Korea prohibited 258 Japanese products, including camcorders and the Sony Walkman, although these restrictions were eased in 1993. South Korean firms attempted, with little success by the early 1990s, to reduce their dependence on Japan by large in-house investments in R&D.

2.10   INVESTMENTS IN THE DRAGONS

The evidence shows that the flying geese model fails to capture the diversity of FDI sources and the manner in which FDI contributes to East Asian development. FDI formed only a tiny proportion of total capital formation in South Korea and Taiwan where successive governments exercised strict controls over the scale and direction of foreign investment. FDI contributed around 2 per cent to South Korea's total capital formation during the period 1976 to 1987 (James 1990 p. 11). In Taiwan from 1965 to 1985 it accounted for between 1.4 per cent and 4.3 per cent of total capital formation and between 2.5 per cent and 5.47 per cent of private capital formation (Dahlman and Sananikone 1990 p. 73).

In the two city states, FDI was more important to economic growth, amounting to roughly 18 per cent of Hong Kong's total capital formation during the period 1976 to 1987 (James 1990 p. 11) and even more to Singapore's capital formation. By 1991 foreign TNCs accounted for around 75 per cent of industrial output in Singapore and around 95 per cent of exports (*Business Week* 30 November 1992 p. 68).

Nevertheless, as the country chapters show, FDI began the production of many fast-growing export lines such as radios in South Korea, Hong Kong and Taiwan. Frequently the TNCs acted as demonstrators and role models for local companies. Some of the larger foreign firms took root and trained local engineers and managers, transferring skills and know-how. Others enabled latecomers to grow through sub-contracting and licensing agreements. Much of the dragons' FDI was concentrated in leading export industries, such as electronics where it helped to create employment and exports. In Taiwan, for instance, FDI contributed between 9 per cent and 16 per cent of total manufacturing employment between 1979 and 1985 and roughly 20 per cent of manufacturing exports between 1974 and 1982.

In the early 1970s, Japanese firms were very active in exploiting the special economic zones set up by the Hong Kong and Singaporean governments. This applied less to South Korea and Taiwan, where restrictions led Japanese companies to favour joint ventures (Urata 1990 p. 5).

Table 2.6 shows the levels of FDI into the four dragons by the two largest investors, Japan and the US. During the entire 1980 to 1988 period, total FDI amounted to only US$14.2 billion, a fairly modest sum when one considers that Samsung's Group sales alone were around US$50 billion in 1991. FDI grew rapidly after 1985, totalling US$11.6 billion, during the three year period 1986 to 1988, compared with only US$2.6 billion over the previous six years. Hong Kong received the largest amount of FDI (US $6.3 billion) over the period. Singapore was second with US$3.6 billion. South Korea received only US$2.3 billion and Taiwan US$2.1 billion. To some extent these figures reflect the openness of each of the economies. Singapore and Hong Kong placed few restrictions on FDI. Conversely, Taiwan and South Korea controlled FDI and protected local industries from foreign competition.

Table 2.6   *Foreign direct investment into the four dragons by Japan and the US (US$ millions)*

	1980–85		1986–8		Total 1980–88		1980–88
	Japan	US	Japan	US	Japan	US	Total (J+US)
South Korea	194	22	1,566	522	1,760	544	2,304
Taiwan	146	114	1,030	761	1,176	875	2,051
Hong Kong	663	552	3,236	1,850	3,899	2,402	6,301
Singapore	491	430	1,543	1,109	2,034	1,539	3,573
Totals	1,494	1,118	7,375	4,242	8,869	5,360	14,229

Sources: Calculated from data in James (1990 p. 15). Original sources: Japan, Export–Import Bank, Kaigai Toshi Kenkyu Shoho, (*Report of the Overseas Investment Research Institution*), November 1988; Ministry of Finance Mimeos, June 1989. United States, Department of Commerce, *Survey of Current Business* November 1984 and August 1988 and 1989 issues.

As a result of rapid growth and some liberalization of foreign investment policies FDI has grown faster since 1988. US net FDI into Asia (excluding Japan), rose by 37 percent from US$1.7 billion in 1991 to US$2.3 billion in 1992 (year-end March). The bulk of US investments went to Singapore and Taiwan. By contrast, Japanese foreign investments amounted to US$5.9 billion in 1992, having fallen from US$7.1 billion in 1991 (Merill Lynch data, Financial Times 16 September 1992 p. 6).

As far as the flying geese model is concerned, Japan was the main investor (with US$8.9 billion) during the 1980s. However, the US invested US$5.4

billion (around 61 per cent of Japanese FDI). Japan was more active than the US in each of the four NIEs and especially in South Korea. However, the US was also a major investor (and supplier of technology), a fact not accounted for in the flying geese model.

2.11 OUTWARD INVESTMENT BY THE FOUR NIES

Comparing Japanese outward investment in East Asia with that of the four dragons shows the growing importance of the NIEs, and hints at the sophistication of many local firms. Unfortunately, accurate, comparable data are scarce. Most data refer to government approvals of investment, rather than actual investment. Table 2.7 presents data on FDI into the ASEAN-4 by Japan and the four dragons for 1988 and 1989. The data show that Taiwan leads the dragons, having invested double that of Hong Kong, its nearest rival, over the two-year period. Outside the ASEAN-4, Taiwan also invested heavily in Mainland China. Gee (1991 p. 26) shows that Taiwan's investment into China grew from around US$100 million in 1987, to US$600 million in 1988 to around US$1 billion in 1989. As noted earlier, Taiwan also overtook Japan as the leading investor in Malaysia in the early 1990s.

Singapore was the third largest investor in 1988 and 1989. South Korea invested least in the region. Although South Korean investment increased markedly in 1989 over 1988, the economy has been relatively slow to relocate production to low cost areas. It is possible that the large size of the *chaebol* made a speedy response to exchange rate appreciations difficult. Also, unlike Taiwan and Hong Kong, South Korea did not have the close cultural proximity with Mainland China.

Within ASEAN the largest recipient of FDI by a wide margin was Thailand, being the most favoured investment location for Japan, Taiwan and Hong Kong in 1988 and 1989. Malaysia and Indonesia followed Thailand. South Korea's most favoured location was Indonesia. Singaporian investment was spread fairly evenly across Malaysia, Indonesia and Thailand. The Philippines attracted far less FDI than each of the other three countries due to political instability and the relatively poorly educated workforce.

The primary motivation for outward investments by the dragons was to circumvent increasing costs at home. Taking Taiwan as an example, outward FDI occurred as a result of exchange rates appreciations, increased foreign exchange reserves and rises in wages. As a result of export success, Taiwan's foreign exchange reserves rose from US$11 billion in 1981 to around US$88 billion in 1992, the largest in the world (*Financial Times* 7 October 1992 p. 6). The New Taiwanese dollar appreciated by around 50 per cent against the US dollar between 1986 and 1989. Overheating of the economy led to

Table 2.7 FDI flows to the ASEAN-4 by Japan and the four dragons 1988 and 1989 (US$ millions)

Investor	Malaysia		Indonesia		Thailand		Philippines		Totals	
	1988	1989	1988	1989	1988	1989	1988	1989	1988	1989
Japan	451	996	226	769	2,827	3,251	95	157	3,599	5,143
S.Korea	15	70	209	466	109	171	2	18	335	725
Taiwan	307	800	929	158	850	867	109	148	2,195	1,973
Singapore	155	338	255	166	275	407	2	24	687	935
Hong Kong	111	130	233	407	445	561	27	132	826	1,230
Sub-totals										
World	1,801	3,205	4,410	4,719	6,233	7,979	452	800	12,896	16,703
J+Dragons	1,039	2,334	1,852	1,966	4,506	5,257	235	479	7,632	10,006
Dragons	588	1,338	1,626	1,197	1,679	2,006	140	322	4,043	4,863

Source: Calculated from data in IEEE (1991 p. 27). Original data from various local sources, compiled by Merill Lynch consultants.

inflationary pressures on wages, land, rents and factory operating costs. The government responded by relaxing foreign exchange controls, enabling locals to invest abroad more easily. The combined effect of these events was to stimulate outward investment into lower-cost zones in the second-tier NIEs and China (especially the nearby city of Xiamen). Between 1987 and 1992, Taiwanese companies invested more than US$3 billion in the Mainland, although no precise figures are available (*Financial Times* 7 October 1992 p. 6).

Hong Kong began investing heavily in China earlier than Taiwan. During the early 1980s, Hong Kong firms began sub-contracting out production in search of cheap labour, as wages rose at home. Hong Kong's investment in China was in the order of US$2.1 billion in 1988, approximately 40 percent of total FDI into China in that year (Chaponniere 1992 p. 122). By 1992 an estimated 3 million Chinese in Guangdong worked in Hong Kong-run factories (*Electronics* April 1992 p. 20).

Regarding the flying geese model, Table 2.7 shows that investment from the four NIEs matched Japan's in 1988 and 1989. In fact, in 1988, FDI by the four together exceeded Japan's by 12 per cent. In 1989 the four NIEs were just behind Japan with an investment of US$4.9 billion (95 per cent of Japan's). In 1990, the four NIEs pulled ahead of Japan in Malaysia, Thailand, the Philippines, Indonesia and China (Tho and Urata 1991 p. 10 and p. 29). Measured on an approval basis, the four NIEs invested US$15 billion compared with Japan's US$6.3 billion for 1990 and US$1.6 billion for the US. The surge of FDI from the four NIEs in 1990 was concentrated in Thailand (US$8.9 billion). Other major recipients were Indonesia (US$2.6 billion) and China (US$2.2 billion). As a group, the dragons have emerged as a highly significant source of FDI for the second-tier NIEs and China.

In short, while Japanese FDI is undoubtedly important to the East Asian region as the flying geese model suggests, the four dragons combined have out-paced Japan both in the second-tier NIEs and in China.

2.12   THE ROLE OF THE DRAGONS IN EAST ASIA

This chapter has shown that the four dragons contributed substantially to trade expansion, economic growth and outward FDI in East Asia, adding their own dynamic to the region. In each of the dragons electronics took off during the 1980s, overtaking other sectors to become the largest export industry overall in East Asia.

Although the flying geese model correctly draws attention to the interdependency of the region, pointing to Japan's technological leadership, the proposition that the four dragons owe their export growth to Japanese TNC subsidiaries is not supported by the evidence. On the contrary, the majority of NIE exports were accounted for by local firms rather than TNCs. Of the four dragons, only Singapore owes its export drive primarily to TNCs, but even in this case, Japanese FDI was exceeded by US FDI and matched by European investment.[14]

The evidence shows that US and European as well as Japanese companies played an important part in the initiation of local industries through FDI, as later chapters demonstrate. To fully understand East Asia's development it is necessary to account for latecomer firms in the NIEs and the competition between Japanese, US and European TNCs in the region.

The flying geese model ignores the importance of the US economy as the dominant export market for East Asian goods. It also understates the weakness of Japan as an importer of NIE goods. Both the US and Europe imported more goods from the NIEs than Japan during the 1980s. In addition, the US has been an important source of technology and skills, both through higher education and training within TNCs. There is little doubt that the US economy provided a technology push as well as a demand pull for East Asian development.

Most importantly for this study, the flying geese model gives little credit to the overseas Chinese as a distinctive force for development in East Asia. The three 'Chinese dragons' did not merely imitate Japan, nor did they depend primarily on Japanese FDI for growth. As Redding (1991) shows, overseas Chinese firms embody their own distinctive styles of management, which reflect long-standing Chinese entrepreneurial skills and customs. The success of overseas Chinese small-scale firms reflects powerful, historical business practices which contrast sharply with Japanese operations within the conglomerate *keiretsu* structures.

Although embryonic, the challenge to Japan is both significant and growing. The aggregate FDI data show that the four NIEs exceed Japanese outward investment in the region and, as the following chapters show, the dynamism of the NIEs is due largely to the skills and tenacity of large numbers of latecomer firms.

NOTES

1. For instance there is very little analysis of intra-regional links, or links with other regions, in the World Bank's (1993) report on the East Asian miracle.
2. For analyses of the Japanese economic and technological systems see Freeman (1987), Odagiri and Goto (1993), Dore (1987) and Okimoto and Rohlen (1988). For high technology in Japan see Okimoto (1989) and for electronics, Gregory (1985). Japanese firm strategies are dealt with by Abbeglen and Stalk (1985).
3. The four second-tier NIEs are also known as the ASEAN-4 economies. ASEAN is the

Association of (six) South East Asian Nations, which includes the ASEAN-4, Brunei and Singapore.

4. Riedel (1988 p. 5) presents data which show that Hong Kong and Taiwan began to grow rapidly during the 1950s. Growth in South Korea and Singapore began to accelerate in the mid-1960s.
5. See Gee (1991) for the causes of Taiwanese inflation. EIU (1990) analyses South Korean inflation.
6. A full analysis of more recent data is provided in each of the country chapters.
7. All official figures in current prices. Note that not all the data are consistent with those in Table 2.4 (e.g. Hong Kong exports are lower than would be expected in 1990). The latter figures are from the consultancy Benn/Elsevier, cited in Ernst and O'Connor (1992). Unfortunately the original sources of their figures are not made clear. The official statistics cited here are probably more accurate.
8. These figures are not strictly comparable for definitional reasons. The Singapore figures include around 25 percent re-exports which is more than the other dragons.
9. Sakong (1993 p. 152) describes Kaname's product cycle version of the model, as well as the work by Saburo Okita. Among the more recent studies which propose the flying geese pattern are Yamashita (1991) and Fukasaku (1991).
10. Note that there are significant differences between the Japanese and South Korean conglomerates, not least in terms of ownership and financial organization (Whitley 1992). For a comparison of Japanese, South Korean and US firm managerial practices, see Chung and Lee (1989).
11. It is not clear whether this figure refers to actual or approved FDI. See Section 2.12 below for official data on outward FDI by the four dragons.
12. By 1991 the Chinese coastal zones together with Taiwan and Hong Kong formed an economic region with a population of 120 million and a combined GNP of approximately US$310 billion, equivalent to Brazil's at that time (Cowley 1991 p. 18).
13. For full details see Chaponniere (1992 p. 73)
14. In 1991, for instance, US firms invested S$962.2 million in Singapore (33.1 per cent of the total), while European firms invested S$684.2 million (23.3 per cent). This compares with Japanese FDI of S$713.2 million (or 24.3 per cent of the total) which is lower than American investment, and only a little more than European FDI (*Singapore Investment News*, February 1992, p. 3).

3.   The latecomer firm

3.1   ANALYTICAL FRAMEWORK

This chapter develops a simple analytical framework for assessing the experiences of the four dragons in electronics in subsequent chapters. A set of arguments are put forward as to the nature of latecomer firms and how East Asian companies were able to close much of the gap in electronics technology. The chapter holds that to overcome their barriers to entry, latecomer firms would need to progress simultaneously on two fronts: export marketing and technology. It is likely that successful companies would pursue strategies to link their export sales to technological learning and innovation.[1] Historically, this is likely to occur through a variety of stages requiring the deployment of various learning mechanisms. The simple model suggests how technological learning occurred as NIE firms graduated from the manufacture of simple goods to the design and development of complex electronics for export markets.

3.2   DEFINING TECHNOLOGY AND LEARNING

Schmooker (1966 p. 18) defined technology as the social pool of the industrial arts. Technology is a resource embodied not only in physical capital but, equally importantly, in human skills, institutions (especially firms) and social structures. Technology represents the capacity to create and extend the existing pool of industrial skills and knowledge. It is not a given or static resource, but rather a dynamic capability used to absorb, adapt and advance existing know-how and skills. Technology acquisition occurs when some or all of the skills needed to absorb and adapt a specific production process or product have been developed.

Technological learning is the process by which firms acquire technology. It is the 'what goes on' in the black box of the conventional (neoclassical) firm of economic theory. Put another way, learning refers to the mechanisms and processes by which technological progress is brought about. As Malerba (1992) shows, learning is central to productivity growth and different types of product and process improvement. Learning is clearly at odds with conventional economic ideas of learning-by-doing which treat technology accumulation as a passive, costless and automatic activity, usually plugged into a production function (Arrow 1962). In contrast with conventional learning-by-doing, technological learning is a dynamic, difficult and costly process. It normally involves substantial and deliberate effort and investment on the part of firms. Learning enables firms to build up their knowledge about products and manufacturing processes, and to develop, deploy and improve the skills of their workforces (Dodgson 1991 p. 23).

Learning is difficult to observe, accurately measure or indeed to distinguish from other manufacturing activities. Corporate learning, as with the learning of individuals, is difficult to analyse, although the outcome of successful learning can be dramatic in effect. The learning process is often a qualitative, informal activity, idiosyncratic in nature, cumulative in effect and uncertain in outcome. Learning usually involves both knowledge and experience, encompassing formal methods such as training and informal mechanisms such as imitation. Learning is usually costly and often difficult to undertake, but it is central to incremental technical change and corporate progress.

This study uses interviews with company directors and engineers to capture the nature of latecomer firm learning and to provide insights into the depth, timing, mechanisms and determinants of learning. The aim is to generate a qualitative understanding of the extent of learning among East Asian firms, over time. The intention is to identify the mechanisms by which firms accumulated technology in electronics, entered international markets and, in some cases, caught up with Western and Japanese market leaders.

3.3   LATECOMERS, LEADERS AND FOLLOWERS

Although learning has been central to technological progress (Fransman and King 1984), there is little analysis of how firms in latecomer countries deploy strategies to learn technology. Since Gerschenkron's classic works on patterns of nineteenth-century European industrialization, many studies have examined the general phenomenon of latecomer industrialization.[2] However, apart from Amsden (1989) and a few other studies, the contribution of firms, their origins, strategies, structures and methods for acquiring technology are rarely treated in the latecomer literature. It is therefore useful to define a latecomer firm and ask how it might develop its marketing and technological capabilities.

For the purposes of this study a latecomer firm is defined as a manufacturing company (existing or potential) which faces two sets of competitive disadvantages in attempting to compete in export markets. The first is technological in character. Located in a developing country, a latecomer firm is dislocated from the main international sources of technology and R&D. It operates in isolation from the world centres of science and innovation and is behind technologically, lacking in research, development and engineering capability. Its surrounding industrial and technological infrastructure is poorly developed. Local universities may also be weak technologically and other educational and technical institutions poorly equipped.

Normally, access to technology and a healthy surrounding national system of innovation is assumed to be essential for corporate competitiveness (Nelson and Rosenberg 1993). To succeed in international markets, the latecomer firm must overcome its technological disadvantages.

The second disadvantage concerns international markets and demanding users. To add to its technological difficulties, the latecomer firm is dislocated from the mainstream international markets it wishes to supply. These are mostly located in the advanced countries, rather than developing countries. Typically the firm will confront underdeveloped, small local markets and unsophisticated users. Many studies show the importance of user–producer linkages and clustering to innovation and industrial development.[3] To succeed, the latecomer firm has to devise ways of overcoming market barriers to entry and then to forge the user–producer linkages which stimulate technological advance. Based in a developing country, the latecomer has to develop outside the major international clusters of innovative suppliers and users.

Latecomer firms are clearly different from leaders. Technology leaders generate new products and processes to gain leadership advantages in the marketplace. Unlike a latecomer, a leader typically has a substantial R&D department capable of generating new innovations and contributing to the firms' competitive advantage. A leader may also enjoy strong and useful connections with universities and other parts of the surrounding technological infrastructure (e.g. IBM and Intel in the US). Through their capabilities, leaders contribute directly to the technological and scientific frontier in their field.

Latecomer firms are also distinct from technology followers.[4] Followers may be behind the leaders, but like leaders they are connected directly into the advanced markets in which they compete. Indeed, in some circumstances, fast followers may have advantages over leaders. They will have substantial technological resources to learn rapidly from the leader's experience, to avoid some of the R&D costs through imitation, and to adapt a product or process more closely to a buyer's need.[5] For example, Fairchild was the leader in the 8-bit microprocessor, but it lost its lead to followers Zilog, Intel and Motorola who introduced improvements to the basic Fairchild design (Langlois et al. 1988 p. 115). As Freeman (1974 p. 176) points out, followers will deploy their R&D resources to profit from the mistakes of others.

Latecomers do not have leader and follower advantages because they are weak technologically and isolated from advanced users. However, latecomers have substantial cost advantages over leaders and followers, and this can form part of their initial market entry strategy. The overall challenge confronting the latecomer is how to devise and implement corporate strategies which enable them to overcome initial market and technological barriers to entry, and then to acquire technology and increase export sales.

3.4   TECHNOLOGY ACQUISITION CHANNELS

Latecomer learning of foreign technology has become embedded in a variety of institutional channels which usually involve foreign firms in contractual arrangements in return for a particular service, such as low-cost production. These channels, presented in Table 3.1 evolved through time as latecomers sought to acquire complex technologies and to compete nearer the technology frontier. The channels, some of which overlap, apply to lesser or greater extents to each of the dragons.[6] Together, they enabled latecomers to acquire technology and enter export markets.

Table 3.1   Mechanisms of foreign technology acquisition by latecomer firms

Foreign direct investment (FDI)
Joint ventures
Licensing
Original equipment manufacture (OEM)
Own-design and manufacture (ODM)
Sub-contracting
Foreign and local buyers
Informal means (overseas training, hiring, returnees)
Overseas acquisitions/equity investments
Strategic partnerships for technology

Source: See text.

FDI and joint ventures were an important starting point for electronics, sparking off new export lines and leading to sub-contracting and OEM. As Schive (1990) and Fok (1991) show, foreign firms acted as demonstrators for local firms to imitate, some assisted local firms to grow through sub-contracting and licensing agreements. Many hired and trained locals in their subsidiaries. While the overall contribution of FDI to capital formation in South

Korea and Taiwan was small (James 1990 p. 11; Dahlman and Sananikone 1990 p. 73), it accounted for a disproportionately large share of electronics exports and employment. In Taiwan the TNCs gave rise to a Schumpeterian process of imitation and swarming on the part of local firms. In some cases, TNCs trained local firms to supply goods under sub-contracting relationships. Several latecomer companies gained direct access to training and engineering support under joint ventures, including Samsung Electronics and Tatung.

Under licensing arrangements latecomers pay for the right to manufacture products usually for the local market, and the TNC transfers the necessary technology for manufacture. Generally, licensing requires more technical capacity than a joint venture where often the senior partner trains the latecomer to manufacture. In Taiwan, between 1952 and 1988 the government approved more than 3,000 licensing agreements (mostly in electronics), many including formal technology transfer clauses (Dahlman and Sananikone 1990 p. 78).

Foreign and local buyers were also an important source of technology and market information in the four NIEs. Hone (1974) shows that many local firms initially sold their goods to large buying houses from Japan and the US. Foreign buyers often placed orders for 60 per cent to 100 per cent of the annual capacity of exporting firms in sectors such as clothing, electronics and plastics. The Japanese buyers (e.g. Mitsubishi, Mitsui, Marubeui-Ida and Nichimen) located in the NIEs to purchase cheap goods as wages rose in Japan in the early 1960s. During the late 1960s they purchased more than US$1.4 billion per annum of low-cost East Asian manufactured goods, 75 per cent of which were sold to the US. This led to a stream of Japanese manufacturers to Taiwan, South Korea and Singapore. Many US retail companies (e.g J.C. Penney, Macy's, Bloomingdales, Marcor and Sears Roebuck) followed suit (Hone 1974 p. 149).

The buyers enabled many firms to expand their production capacity and obtain credit against guaranteed forward export orders. Wortzel and Wortzel (1981), discussed below, show how some NIE exporters progressed from passively selling low-cost production capacity to actively promoting their services to new buyers and setting up marketing offices at home and abroad. Foreign buyers assisted latecomers into export markets and supplied technology in various forms. Often from local offices, they provided latecomers with information on product designs as well as advice on quality and cost accounting procedures. The largest buyers visited factories frequently and supervized the start-up of new operations. Some assisted with the purchase of essential materials, capital goods and components.

A study by Rhee et al. (1984) shows that around 50 per cent of firms in South Korea (from a sample of 113) benefited directly from buyers through plant visits by foreign engineers and visits by Koreans to overseas factories. The buyers provided local companies with blueprints and specifications, information on competing goods and production techniques, as well as feed-back on design, quality and performance. About 75 per cent of firms received assistance with product design, style and detailed specifications. In electron-ics, US retail chains and importers were the most important buyers during the 1970s in South Korea. Buyers helped the latecomers to overcome their dis-tance from the advanced markets and foreign sources of technology.[7]

OEM (a specific form of sub-contracting) evolved out of the joint opera-tions of buyers and latecomer suppliers and became the most important channel for export marketing during the 1980s. Under OEM, the latecomer produces a finished product to the precise specification of a foreign TNC. The foreign firm then markets the product under its own brand name, through its own distribution channels (thereby capturing the post-manufacturing value-added), enabling the latecomer to circumvent the need for investing in mar-keting and distribution. OEM often involves the foreign partner in the selec-tion of capital equipment and the training of managers, engineers and techni-cians as well as advice on production, financing and management. In South Korea, OEM is sometimes linked to licensing deals. Successful OEM ar-rangements often involve a close long-term technological relationship be-tween partner companies, because the TNC depends on the quality, delivery and price of the final output.[8]

OEM is also to be contrasted with own-design manufacture (ODM), first reported by Johnstone (1989 pp. 50–51). The nature and complexity of the OEM system evolved considerably during the early 1980s. According to Samsung, Anam, RJP and other firms analysed in this study, many of the electronic systems purchased under OEM were designed and specified, as well as manufactured, by the local firm rather than the TNC. In 1988 and 1989 this system began to be called ODM in Taiwan. At the time of this research the term ODM was not used by South Korean or Hong Kong firms. However, they too claimed that in many cases equivalent progress had taken place.

Under ODM the latecomer carries out some or all of the product design and process tasks needed to make a product according to a general design layout supplied by the foreign buyer (often a TNC). In some cases the buyer cooperates with the latecomer on the design. In other cases the buyer is presented with a range of finished products to choose from, defined and designed by the latecomer firm with its own knowledge of the international market. The goods are then sold under the TNC's or buyer's brand name as in OEM. ODM signifies the internalization of system design skills, and some-times complex production technologies and component design abilities on the part of the latecomer.

ODM offers a mechanism for latecomer firms to capture more of the value-added while still avoiding the risk of launching own-brand products. Under early forms of OEM, the latecomer was confined to value-added related to assembly services. Under ODM the local company adds value in production engineering and product design. ODM indicates an advance in technological competence, although it is applied mainly to incremental (follower) designs, rather than leadership product innovations based on R&D.

However, as the country chapters show, the OEM system has several disadvantages. Strategically, the latecomer partner is often subordinated to the decisions of the buyer, and often dependent on the foreign company for technology and components as well as market channels. The TNC sometimes imposes restrictions on the activities of the OEM supplier. Without their own distribution outlets, the post-manufacturing value-added is limited. Also the system makes it difficult for local companies to build up the international brand images needed for high quality goods.

Despite the problems inherent with OEM and ODM, it would be wrong to understate the importance of the system. It facilitated rapid industrial growth in electronics and permitted the assimilation of technology. In some cases the more restrictive clauses on OEM and licensing were renegotiated. For example, marketing restrictions on mature products were often set aside so that South Korean firms could sell directly into third countries. The system allowed many companies to achieve economies of scale in production and, in some cases, to justify investments in automation technology. For their part, foreign TNCs continued to benefit from low-cost capacity expansion, enabling rival TNCs to compete with each other. OEM therefore endures as a mutually valued arrangement.

Alongside the formal mechanisms for technology transfer many informal channels exist. These include the hiring of foreign engineers and the recruiting of locals trained in foreign TNCs. Many East Asian engineers went abroad for training in foreign companies, universities, colleges and R&D institutes. As Dahlman and Sananikone (1990) show for Taiwan, informal sources of technology included the copying of products, reverse engineering and the widespread training of foreign engineers abroad. The flow of technically trained Taiwanese returning to Taiwan rose from around 250 in 1985, to 750 in 1989 and more than 1,000 in 1991.[9]

As latecomer firms grew in size and competence, overseas investments became another means of acquiring foreign technology. Companies such as Samsung and Hyundai purchased several high technology firms to acquire skilled engineers and equipment. Strategic partnerships (i.e. joint ventures on a more equal footing) also enabled latecomers to enhance their technological capabilities by developing a new product or process jointly with a foreign company. Samsung, for example, jointed an eight year agreement with Toshiba of Japan in 1992 to develop flash memory chips and with Texas Instruments (TI) to make semiconductors in Portugal in 1993 (*Fortune* 3 May 1993 p. 28).

To sum up, each of the foreign technology channels in Table 3.1 were exploited by latecomer firms to learn skills and overcome barriers to entry into export markets. Most of the mechanisms were dual purpose, providing market and technology access. As the case studies below show, latecomers worked to couple technological and market opportunities, using market signals as a focusing device for technological learning. Over three decades or so, this coupling process has resulted in a substantial, if largely incremental, innovative capacity on the part of many latecomers.

3.5 THE SIMPLE MODEL

As noted above, Wortzel and Wortzel (1981) put forward a simple marketing scheme to show that NIE exporters graduated from supplying labour-intensive assembly services to exporting advanced goods into foreign markets. Their study is based on interviews with locally owned firms in three Asian dragons (South Korea, Taiwan and Hong Kong) as well as Thailand and the Philippines. It is one of the only studies which systematically analyses the export marketing strategies of firms in East Asia. Their study covers three export industries: consumer electronics, athletics footwear and clothing.

Wortzel and Wortzel were not concerned with the technology dimension of firms' development. However, as argued below, it is likely that as firms accumulated marketing skills they also learned technology skills in order to meet increasingly sophisticated customer needs and to capture more of the post-production value-added.

The left-hand column of Table 3.2. summarizes Wortzel and Wortzel's five-stage marketing model. The latecomer firm progressively internalizes the marketing skills and functions initially carried out by the foreign buyer or manufacturer. In the first stage the latecomer is entirely dependent upon buyers for product design skills, marketing, distribution and quality control, while the local firm supplies low-cost production capacity. As the model indicates, during stages two to five the latecomer firm assimilates more and more complex marketing functions. Spurred on by the prospect of growth and profit opportunties, the firm learns how to conduct its own sales and marketing. It progressively broadens its range of customers and improves the packaging and quality of its products.

By stage five the latecomer firm will have developed its own brand design and will organize its own sales either directly to customers overseas or through distributors. No longer is it dependent on the distribution channels of

Table 3.2   Stages of marketing and technology assimilation

	Marketing stages	Technology stages
1.	Passive importer-pull Cheap labour assembly Dependent on buyers for 　distribution	Assembly skills, basic production 　capabilities Mature products
2.	Active sales of capacity Quality and cost-based Foreign buyer dependent	Incremental process changes for 　quality and speed Reverse engineering of products
3.	Advanced production sales Marketing dept. established Starts overseas marketing Markets own designs	Full production skills Process innovation Product design capability
4.	Product marketing push Sells direct to retailers 　and distributors overseas Builds up product range Starts own-brand sales	Begins R&D for products and 　processes Product innovation capabilities
5.	Own-brand push Markets directly to customers Independent distribution 　channels, direct advertising In-house market research	Competitive R&D capabilities R&D linked to market needs Advanced product/process 　innovation

Sources: Technology stages derived from later chapters. Marketing stages summarized from Wortzel and Wortzel (1981).

foreign buyers or manufacturers. In marketing terms, the latecomer will be indistinguishable from leaders and followers.

At the time of Wortzel and Wortzel's research (carried out in the late 1970s) most NIE exporters of electronics had reached stage four (clothing and footwear were further behind). Although many firms had taken control of local marketing, product design and quality, they had yet to establish their own brand names. Wortzel and Wortzel (1981 p. 55) believed that stage five was largely theoretical in the NIEs. By the early 1990s, however, some latecomer firms such as Samsung and Goldstar of South Korea had established well-known brand names in some areas. ACER and Tatung of Taiwan also manufactured own-brand goods. Nevertheless, most latecomers remained behind the leaders, dependent on foreign buyers and TNCs for marketing outlets. For example, Cal-Comp of Taiwan was the world's largest producer of calculators and fax machines in 1991. Although it was virtually unknown in the West, Cal-Comp produced roughly 80 per cent of Japanese Casio calculators under OEM (Cowley 1991 p. 25).

3.6   LATECOMER TECHNOLOGICAL LEARNING

The right-hand column of Table 3.2 adds a technology dimension to the marketing model, suggesting how latecomer firms gradually learn the techniques of manufacturing. The scheme is used to assess the evidence for each of the dragons in subsequent chapters. The model is consistent with a body of industrial research into technological learning in developed and developing countries. Authors such as Dahlman et al. (1985) and Westphal et al. (1985) show that learning tends to develop in sequence, shifting from production to investment learning and ultimately to innovation-based learning. Similarly, Lall (1982) highlights the passage from elementary through intermediate to advanced learning. Lall also points to the importance of non-technological learning to organizational development (e.g. marketing and managerial skills).

However, few if any learning studies systematically relate technological accumulation to exports and export marketing channels, nor do they examine firm strategies towards learning. Usually, the main aim of learning studies is to analyse the general tendencies of learning at the industrial level.

The simple model does not assume any rigid or automatic stages of development. It is a first approximation of how the process of technology assimilation links to marketing development. In the model, progress depends primarily on the strategies and efforts of local firms and the opportunities afforded by foreign buyers. Some of the stages may occur concurrently, as in complex innovation models, and there may be feedback loops between earlier and later stages. External factors (e.g. policy actions and the state of the macroeconomy) will impinge directly on the process. Many of the normal criticisms of innovation models apply to the above scheme (Forrest 1991). The model is consistent with work on international product life cycles (Vernon 1966; 1975) and theories of the location of production (Dunning 1975), although the latter studies are mainly concerned with TNC location decisions rather than the emergence of latecomer firm production.

It is likely that early NIE entrants begin with simple assembly skills and later assimilate incremental process capabilities. As their capacity expands and numbers of customers increase they will need to learn to control the quality and speed of production. During the early stages, firms are likely to remain dependent on outside sources for technology.

Soon, technicians will begin to internalize key production skills. Eventually, the latecomer firm will gain more and more control over its production processes, spurred on by export market opportunities. By acquiring product and process capabilities it will be able to sell higher quality products to a larger base of customers, bringing the advantages of low-cost engineering and management to the market.

The latecomer entrepreneur recognizes that unless the firm goes through a series of difficult technology learning transitions it will remain trapped in the capacity export stage. By stage four, the firm's strategy will be directed to strengthening the skills needed to develop new products and processes. The latecomer will have surmounted its technological dependency in terms of product design, quality control and process engineering. It may have already forged links with capital goods suppliers and may conduct R&D into new products and processes.

The final phase, stage five, is currently of great interest to South Korean and Taiwanese latecomer firms. By reaching this stage they will have developed advanced marketing skills and R&D capabilities, overcoming their latecomer status. At this stage they would be indistinguishable from world market leaders and followers and will be capable of competing at the technology frontier. However, if they fail to progress to these higher stages of product and process development they will presumably retain, at least in some respects, latecomer structures and orientations. This is a theme examined in each of the country chapters.

3.7　LINKING TECHNOLOGY TO THE MARKET

In the simple model, the technological role of exports is to progressively pull the learning of latecomer firms forward by stimulating technological change. Through sub-contracting and other channels, export clients and export demand act as a focusing device for technological investments. Exports force the pace of technological progress and enable latecomers to overcome their distance from the demanding markets of the West. Local competition stimulates the process as export leaders are imitated by followers.

There may not always be systematic, causal links between the stages of technology and market development. It is theoretically possible for a firm to acquire advanced technological skills but still remain at the early stages of marketing – or *vice versa*.

However, it is likely that latecomers will tend to improve both their technology and marketing capabilities simultaneously. Marketing skills are needed for firms to capture the added value associated with packaging, distribution, brand awareness and after-sales service. Marketing capabilities assist firms to expand their range of customers and control the direction of their future business. Similarly, technological know-how is needed to develop new products and improve the efficiency of production. Firms have an incentive to expand their profit opportunities, to manage and reduce their dependency on foreign sources and to respond to competition from other latecomers. These processes require the internal acquisition of both marketing and technological skills.

In some of the stages there may be a direct link between market and technology. For instance, when shifting from stages one to two, firms will need to internalize process skills to expand production capacity, shorten delivery times and improve product quality. To maximize sales of production capacity to key customers, joint engineering work may be needed. Later, to achieve the advanced stage of product marketing push, firms will need sufficient R&D capabilities to convert market signals into innovative new products.

Furthermore, the channels for technology transfer and marketing may be one and the same, as with sub-contracting and OEM/ODM. As the country chapters show, under sales and exporting arrangements latecomer firms are often presented with a technology transfer mechanism. For instance when sub-contracting, the latecomer is often supplied with technical specifications, training and advice on production and management by the TNC. The motive of the TNC is to ensure quality, delivery and price of the final output. These propositions are explored in depth in the individual cases of corporate learning.

3.8   LATECOMERS AND INNOVATION THEORY

According to the simple model above, latecomer firms will tend to enter at the mature, standardized end of the product life cycle. This runs in contrast with the traditional technology life cycle models put forward in evolutionary Western models of innovation.[10] The latter have related empirically observed product life cycles (PLCs) to configurations of product and process technologies (e.g. Utterback and Abernathy 1975; Abernathy et al. 1983). These studies argue that the stage of evolution of a new product is intimately related to process and product innovations.

According to the standard Western model, at the early stage of the PLC, the rate of product innovation is high. Process technology will be relatively experimental and uncoordinated. Products undergo intense innovation, stimulated by market needs. At this stage product markets are ill-defined and nonstandard. As a product is successful in the market place a dominant design (i.e. a standard) emerges, sales increase and market uncertainty diminishes.

Eventually, a product will reach a stage where competition is based largely on price and cost minimization. At this mature phase of the PLC, manufacturing processes tend to become standardized and innovation becomes incremental rather than radical. There is also a shift from the early uncertainty over the technology towards a greater understanding and agreement on market and technological requirements.

Once a dominant design emerges, small uncompetitive firms exit or are acquired by large companies. Eventually, a small number of firms come to dominate the industry by exploiting scale intensive, incremental process improvements. As Utterback and Suarez (1993 pp. 2–3) state: 'Eventually, we believe that the market reaches a point of stability in which there are only a few large firms having standardised or slightly differentiated products and relatively stable sales and market shares, until a major technological discontinuity occurs and starts a new cycle again.'[11]

Although innovation is not a linear process from the R&D laboratory to the market, the creation and diffusion of new technologies are distinct activities: the R&D lab develops, the market selects (Utterback and Abernathy 1975). The outcome of the competitive contest tends to be traceable to the competences, skills and complementary assets that the various rivals bring to the marketplace (Teece 1986). Products and industries undergo life cycles from fluid immaturity states to maturity (Abernathy and Utterback 1979; Kotler 1976). Cycles can cover long periods of gradual evolution, punctuated by short periods of disruptive change (Tushman and Anderson 1986).

3.9   PRODUCT LIFE CYCLES AND LATECOMER FIRMS

The standard model of industrial innovation is intimately linked to the production paradigm of large-volume commodity goods, such as the electronics exports of East Asian firms. However, in contrast with the standard theory, according to the latecomer model above, latecomer firms tend to enter at the mature, standardized end of the PLC and gradually assimilate technology by learning. Latecomer firms are likely to enter at the point where production processes are standardized and cost minimization is paramount. With each wave of new innovations they catch up little by little, closing the technology gap between themselves and the market leaders. In this sense, latecomer firms travel backwards along the PLC. More precisely, with each new wave of product innovations, latecomers move closer and closer to the early activities associated with the early stages of the PLC.[12]

Here, the proposition is that the exporting NIE firm learns its way from the latter stages of technological development to the early stages, working back from the standardized market and technology stages to the more uncertain, early design-intensive and complex innovation stages. During stages one and two, the latecomer will engage with the mature phase of a product's development, focusing on cost minimization as the core competitive strategy. The NIE exporter is able to offer lower costs of production, based primarily on low wage rates. This attracts the attention of buyers, distributors, wholesalers and manufacturing firms in the Western markets, some of whom establish themselves in the NIE.[13]

From the perspective of technological learning, in stage one the NIE firm will begin to build up its basic production capabilities during the course of manufacturing (simple learning by manufacturing). In stage two the company acquires some minor incremental process capabilities, moving backwards along Utterback and Abernathy's model.

The latecomer firm will begin to acquire some control over the quality and speed of production, but is likely to remain dependent on outside sources for process technology. Local technicians are likely to be trained to ensure the internalization of production process technology (learning by incremental process innovation).

The first two market stages of Wortzel and Wortzel correspond to the entry and acquisition of standard process technology. Stage three of Wortzel and Wortzel relates to the more malleable production stages where product design becomes increasingly important. At this stage the latecomer company gains far more control over production process technology. It will have developed considerable product design skills.

In stage four, the firm approaches the product innovation stage. The latecomer will have gained sufficient skills and human resources to develop new products and the new processes needed to make the new products. Strategies will focus on surmounting any remaining technological dependency on outside companies for product design, quality control and process engineering. The firm may well have forged upstream links with capital goods suppliers and may conduct R&D for new products.

Finally, in stage five, the company is as technologically advanced as a leader or follower. The latecomer will carry out R&D for products and processes in competition with Western and Japanese TNCs. Strategies will focus on developing highly complex product and process innovation skills, while proactive R&D will be directed to push the technology frontier forward. By reversing the traditional path of technology development, the firm will have graduated from the mature to early stages overcoming its distance from the world technology frontier. The strategy of linking technology to the market enables the firm to compensate for the lack of demanding local users and consumers.

3.10   INNOVATION, IMITATION AND LEARNING

One of the aims of this study is to assess whether or not latecomer firms
innovate and, if so, in what sense do they innovate. To examine this question
it is first helpful to introduce some basic definitions. Process innovation is
usually defined as a technological change which reduces the cost of making
an existing product or enhances the quality or performance of an existing
product. By contrast, a product innovation involves the development of a
new or improved good.

According to most Western studies, the acid test of an innovation is the
successful introduction of a new or improved product to the marketplace (or
the commercial use of a new manufacturing process) (Dorfman 1987 p. 4;
SPRU 1972 p. 7; Kamien and Schwartz 1982 p. 2). However, this strict
definition, although useful for research into leaders and followers, fails to
capture important industrial and corporate transformations which occur on
the part of latecomers. The latter firms, almost by definition, function from
behind the technology frontier. Therefore, following Myers and Marquis
(1969), Schmookler (1966) and Gerstenfeld and Wortzel (1977 pp. 59–60),
this study defines innovation as a product or process new to the firm, rather
than to the world or marketplace. When a company produces a new good or
service or applies a new method or material it makes a technical change and
an innovation can be said to have occurred. As Myers and Marquis argue,
many firms have been profoundly altered by innovations new to the company, although not new to the world (cited in Gerstenfeld and Wortzel 1977
p. 60). As shown later, this broad definition helps capture the nature of
innovation in East Asia.

In addition, the definition used in this book does not make a stark distinction between innovation and imitation. Imitation is often misinterpreted as
the demonstration of a lack of creativity and talent. However, confirming the
importance of imitation in their study of Japanese industrial development,
Abegglen and Stalk (1985 p. 146) cite from Professor Harvey Brooks of
Harvard University, addressing a US House Subcommittee on Science and
Technology:

> history suggests that imitation, followed by more and more innovative adaptation,
> leading eventually to pioneering, creative innovation forms the natural sequence
> of economic and industrial development. Successful imitation, far from being
> symptomatic of lack of originality as used to be thought, is the first step of
> learning to be creative.

This study therefore pays special attention to learning to imitate as well as
learning to innovate with new products and processes.

Innovation is often a long-term process, rather than a once-and-for-all change. Toyota began their development of the now-famous just-in-time (JIT) car production system in the late 1930s. Under Mr Taiichi Ohno and his associates, during the 1950s and 1960s, Toyota introduced a large number of minor, often difficult improvements (Abegglen and Stalk 1985 pp. 93–104). Yet, it was only in the 1980s that the JIT system reached the eyes and ears of most Western observers.

The JIT innovation example also shows the manner in which innovation is often bound up with non-technological factors. As well as pushing and shoving on a variety of engineering fronts, the JIT system required continuous organizational re-thinking and change. Factory layouts had to be changed. Great efforts had to be made to connect up each of the sub-assembly and fabrication steps, so that material flowed quickly to final assembly lines and inventories could be eliminated.

Another lesson from the Japanese experience is the importance of minor, continuous improvements and the contrast between radical and incremental innovation. A radical product innovation in electronics would, for instance, be the introduction of a brand new, successful product such as the Sony Walkman or the camcorder. By contrast, an incremental innovation would be a minor improvement to the design of a product already on the market. Similarly, a radical process innovation would for instance be the introduction of a new CAD/CAM system. An incremental or minor process innovation would include the modification of production equipment or the improvement of an existing material. In their study of innovation in Japan, Abegglen and Stalk (1985 p. 146) show that Japanese industrial success was based largely on the continuous and creative adaptation of Western, mainly American, technology.

Learning to innovate is therefore a long-term process which cannot be captured solely by counting discrete innovations. In the country case studies careful attention is payed to both technological and organizational innovation, as well as evidence of minor incremental technological change.

3.11   KEY RESEARCH QUESTIONS TO EXPLORE

One major problem in understanding East Asian innovation is the lack of empirical evidence to refute, validate or qualify the model presented above, or any other interpretation for that matter. Some evidence suggests that large buyers assisted latecomer firms to progressively internalize technological skills (Chaponniere and Fouquin 1989; Dahlman and Sananikone 1990). However, there is little analysis of the strategies of individual firms to show how the learning process occurred. Hopefully, the firm-level data on corporate learning paths in the four NIEs will provide general insights into the question of latecomer learning.

One important set of questions concerns the latecomer entrepreneur. What was the origin of the latecomer? How did the firms initially enter international markets? This is an important issue for other developing countries wishing to emulate the success of the dragons. Once the firm entered the export market, how did it build up and strengthen its technological capability? What strategies towards learning and training can be discerned from the evidence on latecomer firms?

Following the logic of the simple model above, other interesting questions present themselves. Do early entrants tend to go through all the stages? Can the stages be carried out simultaneously? As the quality of the local infrastructure improves, can new entrants jump in at the later stages? If so where do they acquire their marketing and technology skills? How do changing factor costs and local market developments affect the strategies of the latecomers? How do firms acquire the skilled engineers, technicians and managers to adapt and improve electronics technology, assuming they do?

Do latecomer firms innovate as they catch up? If so what is the nature and pattern of latecomer innovation? What does innovation mean in the latecomer context? How do firms assimilate innovative capabilities while they are learning? Which factors trigger latecomer innovation? What, if anything, ensures that firms to go beyond keeping up to actually catching up? How do firms change their strategies as they approach the innovation frontiers in the later stages?

At the policy level, it would be useful to know what part East Asian governments played in helping to start up and strengthen their electronics firms. What were the similarities and differences in policy across the four countries and what lessons can be derived from the policy experiences of the NIEs? Which technological infrastructures need to be in place to ensure that widespread catching-up learning can occur?

Finally, at the strategic level, to what extent do latecomer firms represent a challenge to Japan now and in the future? Do latecomers still differ from followers and leaders? What are the current strengths and weaknesses of East Asian firms? Will they be able to overcome their remaining weaknesses to become fully fledged technology leaders?

The rest of the book seeks to answer these questions by examining technological learning at the industrial and firm level in the four NIEs. Illustrative examples of historical learning paths are used to show the patterns of technological transformation as companies moved on to more advanced stages of production. The book explores the learning interface between local firms and foreign buyers and TNCs. Finally, the study attempts to show how the latecomers are changing their strategies to meet new demands as they approach the innovation frontier in electronics.

NOTES

1. In this study strategy refers to the reality of its enactment, rather than in formal pronouncements or statements of intent. As Burgleman and Rosenbloom (1989 p. 19) argue, technology strategy refers to the manner in which technology is sourced, acquired, developed and deployed. The ways in which these tasks are performed contribute to the capability of the firm.

2. See Gerschenkron (1962). A review of latecomer industrialization and Gerschenkron's own contribution is provided by Sylla and Toniolo (1991). Vogel (1991 p. 5) categorizes the four dragons as late late developers. Amsden (1989) uses the latecomer concept to analyse South Korean development and shows how several large companies acquired technology.

3. See, for instance, Marshall's work on industrial districts (1890, ch. 10) discussed by Freeman (1990), Vernon's study on externalities (1960), especially ch. 5 'External Economies', Lundvall's (1988) study on user–producer interaction and Porter's (1990) study on the competitive advantage of nations.

4. As Freeman (1974 p. 269) argues, the idea of leader and follower makes most sense when referring to a firm's strategy in a specific product technology. A large firm may well be a leader in some areas and a follower in others. By contrast, firms from developing countries will tend to be latecomers across a broad range of product technologies.

5. For an analysis of follower and leader strategies see Ansoff and Stewart (1967), Freeman (1974, ch. 8) Porter (1985, ch. 5) and Teece (1986). Freeman's (1974 ch. 8) offensive and defensive strategies are broadly equivalent to the ideas of leadership and followership put forward by Porter and Teece.

6. It is beyond the scope of this study to compare their relative importance through time or across countries, to analyse how the mechanisms relate to each other. Comparing local with foreign sources is also a difficult task as the two are inextricably entwined, with local efforts being essential to absorb foreign technology. Dahlman and Sananikone (1990) provide an in-depth analysis for the case of Taiwan. Schive (1990) deals in depth with FDI in Taiwan.

7. Egan and Mody (1992) and *Forbes* (21 December 1992) show how US manufacturers and distributors purchased bicycles from Taiwan and transferred technology; eventually the Taiwanese succeeded in displacing most of their former American teachers.

8. OEM is similar to sub-contracting in semiconductors and other sectors (e.g. in bicycles and footwear; Egan and Mody 1992). The term OEM originated in the 1950s among computer makers who used sub-contractors (called the OEM) to assemble equipment for them. It was later adopted by US chip companies in the 1960s who used 'OEMs' to assemble and test semiconductors. Since then the term has acquired a variety of meanings. Some use the term to mean the final equipment maker (the TNC buyer), rather than the supplier or subcontractor. To avoid confusion, in this book OEM refers to the system by which firms cooperate in sub-contracting relationships, rather than any particular buyer or supplier.

9. Official statistics cited in *Business Week* (30 November 1992 p. 76). In Taiwan former Bell Labs employees started the Taiwanese Bell Systems Alumni Association which had 120 members in 1992. Similarly in 1994 there were around 80 Bell Labs Alumni in South Korea. Hundreds of others had returned from Caltech, MIT and other leading US technology centres.

10. A critique of the method and data used in the original cross-industry study by Utterback and Abernathy (1975) is provided by Pavitt and Rothwell (1976). A general review of innovation models is provided by Forrest (1991). Metcalfe (1981) also provides useful critiques and extensions.

11. Their study is based on eight industries: manual typewriters, automobiles, electronic calculators, transistors, semiconductors, television sets, television tubes and parallel supercomputers. Note that, apart from supercomputers, these are all high-volume, mass-market industries, where incremental process improvements eventually determine competitive performance. As Pavitt and Rothwell (1976) point out, other types of industry may not follow this pattern.

12. A similar argument to this is put forward by Vernon (1966) for the case of TNCs in developing countries, rather than local firms. The view that latecomers might enter at the standardized end of the PLC is also suggested in a study of innovations made by 33 Taiwanese firms in the mid-1970s by Gerstenfeld and Wortzel (1977 pp. 57–68). This encompassed a range of industries but did not look at progress through time.

13. As Vernon (1966) argues for the case of TNC location decisions, in the case of a mature product and process, the advanced-country firm is freer to seek out lower costs and sources of supply in developing countries.

4. The Republic of Korea: catching up in large corporations

4.1 COMPRESSING THE CYCLE OF TECHNOLOGICAL LEARNING

Just 15 years ago, South Korea[1] had no substantial position in the electronics industry. By the early 1990s, three of the local conglomerates (known as *chaebol*) ranked among the largest electronics producers in the world. As a result of the strategies of the *chaebol* and other fast-growing local companies, production and exports of electronics outstripped steel, automobiles and most other industries by a wide margin during the 1980s.

This chapter explains how South Korean firms gained their competences and overcame barriers to entry, illustrating the forces that shaped and triggered the country's progress. Foreign channels of technology were exploited to the advantage of local firms as they learned production methods, reverse engineered products and accumulated design skills. By linking export demands with in-house investments in know-how, the *chaebol* quickly learned the technology of electronics.

The chapter shows how the government steered early industrial development through its policies and later took a less interventionist role as the *chaebol* grew. Illustrating the achievements of South Korea's latecomer firms in electronics, the chapter charts the growing strength of the *chaebol* through time, pinpointing remaining weaknesses.

Export growth was facilitated by institutional channels, especially OEM, which enabled the latecomers to systematically learn foreign technology. These channels evolved as local corporations compressed the cycle of learning through time.

Individual cases are used to show how latecomer firms transformed themselves into large, competent companies. The chapter illustrates the extent and nature of innovation among local firms, comparing development paths across a range of technologies.[2] Although their skill and tenacity enabled the latecomers to narrow the innovation gap, in some respects they remain latecomers, dependent on foreign competitors for brand names, core components and capital goods. Formulating strategies to overcome these limitations sets the agenda for South Korean firms through the 1990s and beyond.

51

4.2   POLICIES FOR INDUSTRIAL DEVELOPMENT

In the early stages the South Korean Government acted to lead local industry into electronics, delivering the stable macroeconomic environment and the outward-looking industrial approach which enabled the *chaebol* to thrive.

There exists a plethora of studies on the functions of government in promoting industrial development in South Korea (Amsden 1989; Lim 1992; Kim 1989; Kim and Dahlman 1992; Hobday 1991). They show how policies towards the economy, industry and education assisted the growth of the electronics industry.

By fostering the *chaebol* in the 1960s and 1970s, the government influenced the path of technological development in electronics, shaping the overall pattern of industrial structure and ownership in the electronic sector, as it did in other industries. Policies secured favourable exchange rates, high savings, low inflation and low real interest rates: important advantages for local firms. In some cases, targeted industry support projects in electronics may have helped firms climb the technology ladder.[3]

At the outset of industrialization, it should be recalled that per capita incomes were far lower than the other dragons. As market institutions were very poorly developed in the mid-1950s and early 1960s, the government's response was to promote the *chaebol* by procurement, subsidies, protection, tax benefits, credits for R&D and training, technology support and financial incentives.[4] The conglomerates internalized transactions which might otherwise have evolved in a more competitive free market setting.

By contrast, in the 1950s, Taiwan's trading and sub-contracting infrastructure was populated by thousands of traditional Chinese-style family businesses. In 1955 Taiwan's per capita income level was 70 per cent higher than South Korea's (Levy 1988 p. 44). As market failure was widespread, the initial entry costs confronting South Korean firms were high. The government's solution was to foster a small number of large oligopolistic firms with sufficient resources to overcome entry barriers. The Japanese *zaibatsu* (business groups) provided a nearby role model to follow.[5] These policies led directly to the concentrated industrial structures found in South Korea, now among the most concentrated in the world.

Actions to assist the electronics industry included the Basic Plan for Electronics Industry Promotion (1969 to 1976) and the Electronics Industry Promotion Law of 1973. These laws helped set up the black and white TV industry and encouraged industry to acquire technology, to improve output quality and to raise exports.

As part of the fourth five year plan (1977 to 1981) the government: (a) arranged foreign loans of US$221.6 million; (b) established an industrial estate for the production of computers and semiconductors; and (c) set up the

Electronics and Telecommunications Research Institute (ETRI), with a fund of US$60 million (Amsden 1989 pp. 83–4). ETRI later helped Samsung and others to enter the telecommunications industry (Kim et al. 1992; Byun and Ahn 1989).

In 1983 the government protected the local market against competition in computers and peripherals and other low-end electronics, restricting foreign investment in favour of local firms. While the government permitted joint ventures by Hyundai, Daewoo, Goldstar and Samsung with Japanese and US firms (Amsden 1989 pp. 83–4), according to Bloom (1991 p. 9) it also encouraged the Japanese to leave the ROK, once local companies were sufficiently competent. The government withdrew the special tax incentives which earlier helped attract Japanese firms to the free trade zone in Masan and nominated some of Goldstar Electric's factories for military use which led to the withdrawal of NEC from its joint venture with Goldstar.

In an effort to promote exports and forge connections with foreign buyers, in 1962 the government created the Korean Trade Promotion Corporation (KOTRA) which, by the early 1980s, operated about a hundred international trade centres which supplied foreign buyers with contacts, product samples and company information (Rhee et al. 1984 p. 52). Located in the World Trade Centre in Seoul, KOTRA and 30 or so other associations (e.g. the Korea Electronic Product Exporters Association), service industry by tracking markets and supplying information to buyers and exporters.

Competition policies were introduced, not always successfully, to curb the abuse of market power by the *chaebol*, to stimulate efficiency and to prevent monopolistic practices. Despite conflicts with government, as Amsden (1989 pp. 130–36) argues, local firms shared many of the development aims and philosophies of the state. Productivity and competition were enhanced by inter-*chaebol* rivalry, led by powerful business leaders. Today, this rivalry is entrenched in the structures of oligopoly in electronics and other sectors. The origin of the competitiveness of the domestic industry was therefore the direct outcome of successive governments' policies.[6]

4.3   POLICIES FOR EDUCATION AND RESEARCH

At independence in 1945, after 36 years of Japanese colonial rule, nearly 74 per cent of the population of 25 million were illiterate. As few as 300,000 South Koreans had any experience in manufacturing industry as most firms had been run by the Japanese. The economy suffered from a severe shortage of skilled workers, technicians, managers and engineers (Kim 1989 pp. 1–2; Vogel 1991 pp. 48–55).

After the civil war of 1945 to 1953, to improve general education and to increase the supply of technicians, craftsmen and engineers, investment in education grew steadily from 2.5 per cent of the government's budget in 1951 to more than 22 per cent in 1987 (Amsden 1989 pp. 238–9; Kim 1989 p. 18). By the mid-1970s, the illiteracy rate was so small that it was no longer measured. In 1987, 98.8 per cent of children received education up to the age of 14, while college and university attendance grew from around 10 per cent of the population in 1970 to more than 25 per cent in 1987. To add to the 1.4 million students attending higher education in 1988, a further 50,000 studied overseas (*Korea Business World* December 1989 p. 8). By 1990, the country's record on basic education was one of the best in the world (US Department of Education, cited in *Financial Times* 4 May 1990 p. 5).

To compensate for weakness in R&D, large government-funded institutes (GFIs) were set up to carry out applied R&D and engineering, and to train engineers and researchers. One of the architects of the policy argued that the GFIs were needed because in the 1960s the universities had little regard for technology, industry and commerce (interview 1992 with the former Minister for Science and Technology, 1971–5). The Korea Institute for Science and Technology (KIST), set up in 1966, aimed to absorb and adapt foreign technologies and to promote R&D in firms. During the 1970s a further 10 GFIs were created in machinery, electronics, telecommunications, energy and other fields. More than 90 per cent of government research funding was awarded to the GFIs during the 1980s.

Although official statements and academic studies often stress the role of the GFIs, the direct importance of the GFIs to industrial technology is doubtful, particularly through the 1980s. While in the early stage the GFIs may have compensated for a weak academic sector and helped generate manpower, the chief source of learning was the *chaebol* rather than the public sector. As shown below, firms gained their skills through their outward-looking connections with Japanese and Western companies, rather than the GFIs. After the mid-1980s, doubts over the effectiveness of the GFIs led to criticisms of the system. While firms complained that the GFIs lacked industrial relevance, leading scientists argued that they failed to meet top academic standards (Hobday 1991 p. 14; Swinbanks 1993 pp. 377–84).

One of the most important decisions of the Ministry of Science and Technology (MOST) was to lead industrial training and vocational education. At an early stage, the MOST, with the official backing of President Park Chung Hee, invested in craft and engineering programmes. In 1973 a new law decreed that craftsmen had the same status as scientists and engineers; each year the President would award a prize to all three on the same platform, where previously only top academics had stood.

Such symbolic measures were coupled with creative practical measures, including military exemptions (seen as one of the highest awards in the ROK) for students in engineering and craft who managed to obtain scholarships. While South Korea encouraged craft and engineering apprenticeships, few other non-East-Asian developing countries took craft and technician skills so seriously.

4.4   STAGES OF TECHNOLOGICAL DEVELOPMENT

Technology and Economic Growth

A remarkable economic performance by any standards, South Korea's GNP rose from US$87 per capita in 1962 to US$243 in 1970, to US$1,481 in 1980, to US$6,749 in 1992 (EPB data cited in Kim 1994 pp. 12–13). Exports grew from US$87 million in 1962 to US$76.6 billion in 1992. By 1992 just over 70 per cent of exports were accounted for by industrial goods including electronics, machinery and chemical products, while the share of agriculture in GNP declined to 7.6 per cent in 1992.

Technology was a vital factor in economic growth. Many studies show the country's vigorous advance in technological and human resources over the past three decades.[7] However, despite increased government spending, the state's importance to technology has declined in favour of the *chaebol*, especially during the 1980s.

Figure 4.1 characterizes the stages of South Korean development, using R&D manpower supply as an indicator of technological investment and output. By 1990 R&D manpower had risen to a total of 70,503 of which 38,737 (55 per cent) worked in private R&D laboratories, 21,332 (30 per cent) in educational institutes and 10,434 (15 per cent) in government research institutes (MOST 1993 p. 23).

Most other technology and innovation indicators (including patent outputs, publications, R&D manpower per 1,000 population, corporate R&D spending, numbers of private R&D laboratories and educational expenditures) show a similar pattern: a slow start-up in the 1960s, a take-off in the 1970s, followed by rapid growth through the 1980s and into the 1990s. For example, following a take-off period in the 1970s, government S&T spending grew more than seven-fold from US$702 million in 1981 to US$5,259 million in 1990 in current prices, and nearly five-fold in real terms, increasing as a percentage of GNP from 0.89 per cent to 2.2 per cent (MOST 1993 p. 21).

Technological progress mirrored the stages of economic development. After post-war reconstruction, assembly of shoes, clothing, textiles and light goods emerged in the 1960s. Intermediate goods were also established, often

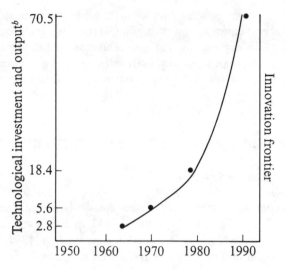

Stages of development:

1950s	1960s	1970s	1980s	1990s
Aid, labour intensive production	Imitation Simple techn. Import of old plant/machinery	Heavy industry Assimilation Minor innovations	Formal R&D Process adaptations New product Dev. Electronics	R&D-intensive Increase in science Fundamental R. New product innov.

Notes:
[a] No linear progression is implied, but a general tendency for firms and industry to move from simple to complex technological activities through time.
[b] R&D manpower (in thousands) supply is used here as an indicator for technological investment and output (see text for other measures).

Figure 4.1   Stages of technological development in South Korea[a]

relying on technology imports. During the 1970s, the heavy and chemical export industries were mobilized along with steel, construction and ship building. Technical education was expanded and foreign technology was adapted and improved upon by local firms. In the 1970s the leading *chaebol* increasingly looked to electronics as a new export growth area (see below).

The Declining Importance of Government

Figure 4.1 illustrates the progress of South Korean firms towards the innovation frontier (the right-hand vertical axis), defined as the point at which R&D for new products and processes becomes central to competitiveness. Leaders such as Intel, AT&T, IBM and NEC compete at the innovation frontier, developing next-generation products and investing heavily in basic R&D.

Through time the *chaebol* narrowed the gap between themselves and the leaders by investing in technology.

Taking control of R&D, by 1993 private firms accounted for over 80 per cent of total R&D spending, up from around 48 per cent in 1980, and just 10 per cent in 1965 (Swinbanks 1993 p. 377; Lim 1992 p. 14). As an example, Samsung's Advanced Institute of Technology (SAIT), established in the late 1980s as the Group's central laboratory, employed around 450 researchers and 150 support staff in 1993, spending around US$50 million per annum (much more than most government institutes). In all, Samsung's 26 companies boasted around 1,000 researchers, plus 13,000 development engineers. Senior Samsung staff travelled regularly to the US and Japan on recruitment drives. Goldstar too built an impressive central research laboratory, employing some 200 staff headed up by an ex-employee of Digital Equipment Corporation (DEC) in the US.

Through their behind the frontier catch-up strategies, the *chaebol* narrowed the gap between themselves and the market leaders through a painstaking, incremental process which accelerated rapidly during the 1980s. As shown below, much corporate R&D was devoted to acquiring and assimilating foreign technology, while innovation was concerned with continuous improvements to processes and product designs, rather than the generation of radical new products through basic research.

Despite R&D spending, licensing and sub-contracting were still important sources of technology in the early 1990s. Often, local designs applied to low-end goods such as Hyundai's 1994 Accent Model (the successor to the Excel Model), a car developed from design to assembly entirely with Hyundai's own technology (*Business Korea* March 1994 p. 12). But in high-end products and complex electronics (such as microprocessors, application-specific integrated circuits and workstations), South Korea lagged behind the market leaders.

4.5   PHASES OF DEVELOPMENT IN ELECTRONICS

Historical Performance

South Korea's electronics industry began with the manufacture of transistor radios for the home market in 1958. In the 1960s American and Japanese firms invested in assembly activities in search of cheap labour. As the industry grew, the share of electronics in manufacturing rose from 2.1 per cent in 1970 to 6 per cent in 1980 to 17.8 per cent in 1988. As Table 4.1 shows, exports of electronics rose steadily to reach US$2 billion in 1980, around 11.4 per cent of total exports. During the 1980s electronics exports increased

Table 4.1   *Exports of electronics goods by South Korea 1961 to 1991*
              (selected years) (US$ millions)

Year	Electronics (US$)	Share in total exports (%)	Rank
1960	0	0	–
1970	29	3.5	5
1975	453	8.9	2
1980	2,004	11.4	2
1985	4,285	14.1	3
1988	15,200	25	n/a
1991	20,157	28	1

Sources: Korea Foreign Trade Association, *Trade Yearbooks*, cited in Sakong (1993 p. 232); 1988 data from Jun and Kim (1990 p. 3).

ten-fold in current dollar terms. By 1991 electronics had become the largest export, accounting for some US$20.2 billion, or 28 per cent of total exports. Other important exports (in order of size) were textiles and garments (US$15.5 billion), steel (US$4.5 billion), shipping (US$4.1 billion) and automobiles (US$2.3 billion) (Sakong 1993 p. 232).

Through the 1980s, electronics were the centrepiece of the economy's export-led industrialization. Product lines included colour TVs, semiconductors, videocassette recorders, computers and peripherals and fax machines. In 1994 semiconductors became the largest single export, amounting to an estimated US$8.4 billion (*Business Korea* March 1994 p. 24).

Phase 1: TNC-dominated development

Progress in electronics can be divided into three broad phases. Phase 1, which ran from around 1960 to 1970, was dominated by FDI. Some exports of domestically made transistor radios started in 1962 (Suh 1975 p. 109). US firms including Motorola, Signetics and Fairchild began to assemble chips during the mid-1960s. They were followed by Japanese and joint Korean–Japanese ventures such as Samsung–Sanyo, Crown Radio Corp, Toshiba and Goldstar–Alps Electronics (Suh 1974 pp. 17–19). In 1972 the largest local firm was Goldstar, followed by smaller companies such as Tai Han Electric Wire and Ho Nam Electric Company.

During the mid-1960s, as a strategic export industry, the government offered TNCs incentives and privileges. As a result, electronics exports rose from around 17 per cent of production in 1965 to 74 per cent in 1973 (Suh 1975 p. 109). In 1972 eight large foreign companies accounted for some 54 per cent of total electronics exports and around 34 per cent of production. At that time there were 253 (mostly small) local firms, as well as 27 joint ventures. While foreign firms specialized in the assembly of parts and components, local companies assembled radios and black and white TVs for the protected local market and later on for export.

Phase 2: local firms and joint ventures

Phase 2 (*circa* 1970 to 1979) saw more local firms and joint ventures assisting in the take-off stage for electronics. Production increased from US$45.9 million in 1968 to US$3.3 billion in 1979, while exports grew from US$20 million to US$1.8 billion. In 1979 parts and components accounted for 48 per cent of total production (mostly integrated circuits, condensers and transistors), 42 per cent was consumer electronics (led by black and white TVs, tape recorders and amplifiers), while 10 per cent was industrial equipment (mainly telephone exchanges, desk-top calculators and transceivers). Exports followed roughly the same pattern as total production.[8]

The components sector was composed mainly of fairly low technology, semi-finished parts shipped into South Korea for final assembly by foreign firms and joint ventures, exploiting low domestic wages. Soon after one or two large TNCs started assembly in South Korea other firms followed to reduce their own costs.

During the 1970s a substantial increase in the share of local and joint venture firms in total output occurred. By 1979 foreign companies accounted for around 40 per cent of exports, down from 71 per cent in 1968. Joint ventures accounted for 15 per cent in 1979, compared with 8 per cent in 1989. Local companies increased their share of exports from 21 per cent in 1968 to 40 per cent in 1979.

The latecomer industry was still fairly dispersed and weak in the mid-1970s. Of the 691 firms registered as electronics producers in 1977, 480 were Korean-owned (mostly small companies), 167 were joint ventures and 44 were foreign-owned ventures. Small local enterprises produced labour-intensive, simple goods for the local market and provided services for the larger companies. Of the small firms, 40 produced capacitors while 200 or so made printed circuit boards.[9]

The US was the single largest electronics market during the 1960s and 1970s, absorbing around 43 per cent of the country's exports in 1979, compared with 17.4 per cent for Japan, 7.8 per cent for Hong Kong, 7.2 per cent for West Germany, 2.4 per cent for Great Britain and 2 per cent for France.[10] This followed the typical pattern, with Japan lagging behind the US as an importer of East Asian goods, but enjoying large export surpluses with each of the dragons.

Phase 3: growth and sophistication of the *chaebol*

The third phase, which took place during the 1980s, saw rapid market growth and the *chaebol* becoming larger and more sophisticated as they took over from foreign firms, not only in exports of consumer electronics but also in industrial and professional goods. Telecommunications output, which accelerated after an easing of regulations on foreign technology imports in 1984, grew from US$458 million in 1983 to around US$1.1 billion in 1987.[11] Most equipment was sold domestically to upgrade the local network. PC production, mainly for export, grew from just over US$50 million in 1983 to US$444 million in 1987, while peripherals soared from US$136 million to US$912 million.

Consumer goods production, which rose from US$3.3 billion to US$7 billion between 1983 and 1987 (a compound annual growth rate of 26.9 per cent), was exported mainly to the US. South Korean exports of colour TVs, VCRs, audio equipment and microwave ovens benefited from the appreciation of the Japanese Yen in 1985.

Similarly, output of parts and components grew from US$2.4 billion in 1983 to US$7.3 billion in 1987 (an annual average compound rate of 31.8 per cent). However, in this area foreign firms accounted for 24.3 per cent of exports in 1987, compared with domestic firms at 44.2 per cent and joint ventures 31.5 per cent. In that year the largest component exports were semiconductors (US$2 billion), video tapes (US$352 million), cathode ray tubes (US$314 million) and magnetic heads (US$176 million). Capacitors, tuners, motors and printed circuit boards were also significant.

During the third phase, the *chaebol* increased their overseas manufacturing as they invested close to customers to avoid trade restrictions. Total outward investment rose from an accumulated US$476 million in 1985 to US$1.1 billion in 1988. By 1990 the figure exceeded US$2 billion (Hobday 1991 p. 39). Samsung, the leading investor, began making colour TVs in the US and Portugal in 1982, semiconductors in the US in 1987, microwaves in the UK in 1987, new media products in Canada in 1989, VCRs in Spain in 1990 and telecommunications and TVs in Hungary in 1990 (Hobday 1991 p. 40). Hyundai set up an automotive plant in Canada in the late 1980s.

During the three phases of growth, South Korean latecomers diversified, became more sophisticated and narrowed the technology gap between themselves and the overseas market leaders. Firms progressed from simple radio and TV assembly in the 1960s to colour TVs, colour monitors and advanced semiconductors in the late 1980s.

Structure of Production: Early 1990s

Electronics, the largest industry in South Korea, accounted for nearly 30 per cent of total exports in 1992, amounting to Won 26.2 trillion (around US$33.4 billion) according to the Ministry of Trade, Industry and Energy. Table 4.2 shows the structure of production for 1992.

Table 4.2   Structure of production in electronics 1992

1. Semiconductors and other components	Won 12.3 trillion (US$15.7 billion)
2. Household appliances	Won 8.2 trillion (US$10.4 billion)
3. Industrial electronics	Won 5.6 trillion (US$7.1 billion)

Source: *Korea Times* 19 March 1993 p. 8 (calculated on the basis of current exchange rate).

With rising wages, the domestic market absorbed more electronic output. By 1992 around US$12 billion (or 36 per cent) of production was for local consumption, while US$21.6 billion (64 per cent) was exported. Despite the slump in the world market, South Korean electronics exports rose by 7 per cent in 1992 making it the sixth largest producer in the world after the US, Japan, Germany, France and the UK.

Official data for 1991 show export specializations and strengths.[12] Of the total exports of roughly US$19.1 billion in 1991, parts and components constituted US$9.9 billion, consumer equipment US$6.1 billion and industrial equipment US$3.9 billion. The largest single item was semiconductors (US$5.6 billion). Exports of video equipment were US$2.9 billion, of which US$1.5 billion was colour TVs (black and white TVs had fallen back to US$0.1 billion). Also exported in 1991 were US$1.3 billion worth of video-cassette recorders, US$1.2 billion of cathode ray tubes (mostly colour monitors), US$0.9 billion of magnetic tapes and parts and US$0.6 billion of video tapes.

Other important exports in 1991 were US$2.5 billion worth of computers and peripherals, made up of US$1.8 billion of peripherals and US$0.7 billion in PCs and other computers (mostly 16 bit). Total telecommunications production amounted to US$5.1 billion but only US$23.2 million were exported, the balance was used in the local network.

4.6   STRENGTHS AND WEAKNESSES IN ELECTRONICS

Strengths

The data show that by the early 1990s local exporters had strengths in large volume, medium level technology, rather than leading edge or niche markets. Consumer electronics typically followed the lead set by Japan by one or two years. This gap, much wider during the 1960s, 1970s and 1980s, had narrowed considerably as latecomers caught up.

In some areas (e.g. semiconductors) South Korean companies were ahead of many foreign TNCs in both production and design technology. Samsung's engineers astonished observers in July 1990 when they presented samples of 16 megabit dynamic random access memory (DRAM) semiconductors to conferences in the US, ahead of the market leaders in Japan and the US, not to mention Europe. Ten years earlier, Samsung had no significant market share in chips. By 1990 it had become a market leader in one of the most challenging areas of electronics.

In 1992 local firms commanded a 12.1 per cent share of the world market for memory chips, including 19.7 per cent of the (then) mainstream technology, the four megabit DRAM. Samsung was ranked fifth among the world's DRAM producers and first in one and four megabit DRAMs (Dataquest Inc. data cited in *Business Week* 30 November 1992 pp. 70–71). Other South Korean firms were also significant players in the DRAM market. Goldstar Electron and Hyundai Electronics Industries ranked twelfth and thirteenth in the world in 1992. Samsung and Goldstar both had plans to invest around US$1 billion in new 16 megabit DRAM manufacturing facilities in 1993. In 1993 Samsung announced its own version of the latest 64 megabit DRAM, again ahead of Japanese and US firms. Reflecting its own competence, it formed an eight year partnership with Toshiba of Japan to develop flash memories, an important new type of memory chip technology (*Far Eastern Economic Review* 7 January 1993 p. 57).

Similar indicators of South Korean achievements in world markets abound. To overcome their dependency on licensing, local firms increased their R&D efforts, acquired overseas firms and formed technology partnerships with many leading foreign companies. Samsung led the other latecomers with investments in Silicon Valley and elsewhere. In 1991 the Group invested around 9 per cent of its total sales in R&D, in line with the spending of the leading Japanese corporations (Koh 1992 pp. 28–9).

Weaknesses

In spite of these achievements, the industry suffered from persistant structural weaknesses in the early 1990s. The advances noted above only applied to a small range of technology areas and a small number of firms. Most production was in relatively mature components, and large volume, low-end, low-cost consumer and industrial electronics. Much of South Korean production was behind the technology frontier, produced under OEM, sub-contracting and licensing arrangements.

Under OEM, the most important outlet for consumer goods, foreign companies captured much of the post-manufacturing value-added. Often the latecomers relied on foreign companies (sometimes their competitors) for key components, materials, designs and capital goods.

Technology imports forced the economy into consistent, heavy trade deficits with Japan throughout the 1980s and the early 1990s. The large deficits with Japan were usually balanced out by equally large export surpluses with the US and Europe (Chaponniere 1992 p. 73). In 1991, imports of around US$21 billions in high technology industrial goods from Japan produced a bilateral trade deficit of US$8.8 billion (Holden and Nakarmi 1992 pp. 24–5). In 1991 alone imports related to electronics amounted to US$11.2 billion.

OEM, licensing and sub-contracting limited the strategic scope of local companies in world markets, often placed restrictions on marketing efforts and forced latecomers to rely on Japanese and US leaders for distribution outlets. Such arrangements subordinated the *chaebol* to the strategies of their senior partners and would-be competitors.

Throughout the 1980s, Samsung and the others made steady progress in own-brand sales by advertising and own-product developments, but most remained dependent on foreign firms for key components and licensing. Goldstar licensed the 4 and 16 megabit DRAMs from Hitachi of Japan; Hyundai also relied on licensing from Japanese firms. While Samsung had managed to develop its own DRAM technology in-house, in complex areas such as reduced instruction set computer (RISC) microprocessors, it licensed in from Hewlett Packard (HP) of the US. Samsung was also weak in applications-specific integrated circuits (ASICs) and static random access memories (SRAMs). Overall, the ROK chip industry imported nearly 75 per cent of all the materials and equipment needed for production. In computers, royalties to Intel, Microsoft and others accounted for up to 10 per cent of corporate sales (*Business Korea* March 1994 p. 25).

By the late 1980s lower-wage East Asian countries (e.g. China, Malaysia, Indonesia and Thailand) began to undercut South Korean firms. Responding to higher domestic costs, the *chaebol* relocated some production to lower-cost countries and concentrated on higher-end, higher-quality goods, at home.

Employment fell in labour-intensive areas such as garments, footwear and low-end electronics in favour of engineering-intensive goods such as hard disk drives, complex peripherals and PCs.

Problems also confronted the *chaebol* at the more innovative, research end of electronics. Although design and R&D capabilities had improved, compared with the developed countries, the South Koreans remained weak. The number of research scientists and engineers per 10,000 workers in 1990 was around 33, far behind Japan's 87, the US's 77, Germany's 56 and Switzerland's 44. The ROK was ahead of Singapore with 28, but behind Taiwan with 43 (NSTB 1991 pp. 12–17). R&D spending as a percentage of GDP in South Korea lagged behind most developed countries (MOST 1993 p. 37).

The fragility of the science base was highlighted by the number of South Korean articles published in international academic journals. The Science Citation Index (SCI) showed that in 1990 the ROK ranked only thirty-second in the world, with only 1,800 publications, compared with nearly 250,000 for the US and nearly 50,000 for Japan. According to the SCI quality index (average citings per paper) the ROK came next to last in South East Asia and far behind the developed countries (Swinbanks 1993 p. 379).

Another weakness resulted from the vertical integration strategies of the *chaebol*. Because they internalized so many of their activities, the small and medium sized enterprise (SME) sector was underdeveloped. Compared with Japan, where large numbers of SMEs assist large firms with services and technology, most South Korean SMEs were low-technology, low-cost producers, not yet capable of playing a very dynamic role.

To sum up, despite its successes, significant structural weaknesses faced the electronics sector in the early 1990s: dependence on Japan for core components and market outlets; competition from lower cost economies; the weakness of the S&T base; and the lack of a thriving SME sector. Except in a small number of areas, the *chaebol* were, as yet, unable to compete at the technology frontier.

4.7   INSTITUTIONS FOR LEARNING FOREIGN TECHNOLOGY

Dual-Purpose Market and Technology Channels

Over the past 30 years or so a variety of institutional channels, usually involving foreign firms in transferring technology for a payment or service, enabled companies to learn foreign technology. Often the institutional mechanism for learning was precisely the same as the mechanism for export marketing.

This point is important, as usually technology is analysed independently of market developments. As all the country chapters show, latecomer firms coupled together technology and market mechanisms, enabling them to use the export market as a focusing device for technological investments. Over three decades this process resulted in the fine tuning of local innovation to the needs of the export market. Today this is a commonly admired feature of East Asian companies.

FDI and joint ventures enabled local firms to make simple electronics. Licensing tended to be used once firms had acquired sufficient in-house skills to operate at arms length. OEM proved to be the most enduring export marketing/technology transfer channel in consumer electronics. With sophistication, overseas acquisitions and strategic partnerships enabled some of the latecomers to extend their global technological reach.

FDI and Joint Ventures

During the mid-1960s US firms entered South Korea to assemble products using cheap labour. According to Bloom (1991 p. 10), these TNCs imparted little know-how, importing most of their inputs. However, they began the electronics industry by demonstrating South Korea's low-cost labour and manufacturing. Many South Korean firms imitated US companies, as did Japanese firms mostly through joint ventures.

Table 4.3 shows how leading Japanese firms formed joint ventures in the late 1960s and early 1970s with South Korean junior partners. Samsung and the others diversified into electronics from other industrial areas. They offered low-cost finance and labour in exchange for know-how and export outlets. By 1976, around 50 per cent of electronics employment was accounted for by foreign-owned or joint venture firms (Bloom 1991 p. 9). Joint ventures first applied to consumer goods and later to telecommunications. Samsung Electronics began as a joint venture with Sanyo of Japan in 1969. Its first step was to acquire overseas training, machinery, components, raw materials and foreign management techniques from Sanyo. As wages rose in Japan, Samsung offered Japanese companies capacity for producing large-scale, low-cost, standardized goods. South Korean overhead costs were pared to an absolute minimum to meet Japanese demands.

Samsung sent engineers and managers to firms in Japan, the US and Europe for training. For example, in the early 1980s the company despatched a hundred or so young engineers to Belgium for training in telecommunication switching systems, under a licensing deal with ITT/Alcatel (interview Samsung 1993). Later in the 1980s Samsung formed a joint venture with the US firm GTE to gain telecommunications know-how.

Table 4.3   Early joint ventures and technical assistance from Japan

Year	Technology activity and firms
1961 and 1962	Matsushita and Sanyo provided technical assistance to Samsung and Goldstar to set up transistor radio factories in 1961 and 1962
Late 1960s	Toshiba formed a joint venture and two major technical agreements to assemble consumer goods, cathode ray tubes (CRTs) and parts for CRTs, in South Korea
1969	Samsung Electronics Engineering began assembly of black and white TVs following technology transfer agreements with Sanyo
1969	Joint venture (Samsung–Sanyo) formed to manufacture electronic parts
1970	NEC formed two joint ventures, with Goldstar Electric and Samsung. Samsung–NEC produced electronic components for CRTs
1973	Samsung Electronic Parts, a joint venture, was established in 1973 with Sanyo
1973	Anam Industrial of South Korea formed a jointly-owned venture with Matsushita of Japan to produce colour TVs
1973	Samsung joined Corning of America to acquire technology to produce the glass for CRTs

Sources: Compiled from Bloom (1991 p. 8); Archambault (1992 p. 8).

4.8   THE OEM AND ODM SYSTEMS

OEM: a Technology Training School

During the 1980s, the share of foreign ownership in electronics fell considerably. Despite growth, employees in foreign-owned plants fell by one third between 1976 and 1985. Japanese TNCs including Matsushita, Sanyo and

NEC withdrew from joint ventures as tax advantages were cancelled and firms were encouraged to leave by the government (Bloom 1991 p. 9).

With the growing competence of local firms, the OEM system provided an alternative to joint ventures. OEM enabled South Korean companies to export huge volumes of goods under foreign brand names and distribution channels. Under OEM, often linked to licensing and joint ventures, TNCs helped train engineers, select equipment and supply materials and capital goods to the latecomers.

In consumer electronics, computers and microwave ovens, OEM was a harsh industrial training school for South Korean firms who learned their skills from Japanese and American TNCs.[13] Invariably, production had to be to the highest quality at the lowest price. As the case of microwave ovens shows, to gain its first export order from GTE, Samsung trimmed costs to a minimum, while production engineers worked long hours for seven days a week, sometimes sleeping by their machinery (Magaziner and Patinkin 1989). Samsung and others invested heavily for little or no return, just to win their first small export orders; the TNCs gained the large majority of the profits. If one firm failed to meet expectations, the buyer could switch to another eager South Korean (or Taiwanese) supplier.

Within the OEM channel, South Korean firms learned by strenuous in-house efforts, by trial-and-error investments and by on-the-job training. Eventually they mastered much of the production and design know-how for electronics, narrowing the gap with the leaders. OEM allowed more firms to expand into more areas, to overcome barriers to entry and to access Japanese-controlled channels of distribution into Western markets.

Japanese and US firms sponsored latecomers in this way in order to gain from rapid low-cost expansion of manufacturing capacity. OEM enabled market leaders to compete with each other by reducing production costs, particularly after the Yen appreciated in the mid-1980s. Once one market leader began OEM, others quickly followed or suffered the consequences. In the case of microwave ovens, Samsung may well have prevented the US firm GTE from exiting the market under the pressure of Japanese competition (Magaziner and Patinkin 1989).

Dependence on OEM and Licensing

Confirming the importance of OEM, Jun and Kim (1990 p. 14) estimate that in 1988 OEM accounted for around 60 per cent to 70 per cent of total electronics exports. *Electronic Business* (22 April 1991 p. 59) put the figure at 70 per cent to 80 per cent of total ROK electronics exports in 1990, estimating that the largest three *chaebol* relied on OEM for 60 per cent of their exports. Similarly, a survey conducted in 1993 by the Korean Foreign Trade Association (KFTA) showed that 61 per cent of all exports (including non-electronics) to Europe were conducted on an OEM basis. Very few firms had established their own sales outlets into Europe (*Korea Times* 12 March 1993 p. 9).

In electronics, the OEM system evolved as latecomer firms began to supply more production and design technology, but they remained dependent on foreign firms for key components and capital goods. OEM and licensing were particularly important for products new to South Korean firms (e.g. advanced computer terminals, large telecommunications exchanges and semiconductors). As products matured and capabilities were learned, OEM became less important (e.g. in audio and TV).

Under licensing deals, often linked to OEM, South Korean companies payed royalties for patent rights, as well as product, process and components technologies. Samsung, for example, in the early to mid-1980s formed licensing deals with Toshiba, JVC and Sony of Japan and Philips of Holland (Koh 1992 p. 40). Products licensed included colour TVs, videocassette recorders, air-conditioning equipment and microwave ovens. Licensing enabled firms to acquire telecommunications technology from Alcatel, Toshiba, Italtel and GTE in the mid-1980s, allowing the *chaebol* to master production techniques and to learn to design new products (see Section 4.13).

As noted, reliance on OEM brought with it considerable difficulties, limiting suppliers to low-quality product ranges and forcing latecomers to depend on the strategic decisions of rival Japanese firms. South Korean managers worried that once South Korean firms reached parity in new product areas, Japanese firms would restrict the supply of components and capital goods (interviews with Samsung and Hyundai, 1993). South Korean currency appreciations and wage rises in the latter half of the 1980s made OEM less attractive to Japanese firms, who went in search of lower-cost sites. The lack of high-quality international brand images was felt by many South Korean companies to be a long-term constraint on growth.

To create brand awareness abroad, the three largest *chaebol* advertised in most of their main markets through TV, billboard posters, airport trolleys and so on. They organized distribution facilities close to customers overseas, sometimes risking large investments. In computers Hyundai attempted, with little success, to market its PCs directly to customers in the US. Although its PC revenues had reached around US$228 million in 1991, few US buyers recognized the Hyundai brand name, while US distributors became irritated at constant design changes, unreliable customer service and late shipments. To improve its US position the company launched a US$14 million (per annum) marketing programme in 1992, adding a further US$10 million to print advertising (*Electronic Business* 16 February 1993 pp. 61–4). The risk and difficulty of dispensing with better-known foreign brand names was also felt by many Taiwanese firms (Chapter 5).

The Shift to ODM in South Korea

Technological activities under OEM evolved considerably during the 1980s, according to senior executives of Samsung and other firms. By the late 1980s, many of the goods purchased were designed and specified, as well as manufactured, by the local firm, while the foreign buyer simply branded the ready-made product. In Taiwan this began to be called ODM in the late 1980s. During interviews with many engineers, it became clear that, although the term ODM was not used in South Korea, a shift from OEM to ODM had taken place in a variety of areas.

Under ODM the South Korean companies carried out some or all of the product and process tasks, according to a general design layout provided by the TNC or buyer. Some ODM purchases involved close design partnerships, while others were at arms length, involving little input from the buyer. As with OEM, goods were then sold under the foreign buyer's brand name. ODM signified more advanced design skills and often complex new production technologies. Internalization of such activities marked a further increase in the innovative capacity of latecomers. ODM enabled ROK firms to gain more of the total value-added, while avoiding the expense of investing in distribution channels. Although indicating more capability, it applied mainly to follower, incremental designs, rather than new product leadership. As of the early 1990s, OEM and ODM remained a thriving technology-market channel for many South Korean companies.

Strategic Partnerships and Enhanced Domestic R&D

As large firms approached the technology frontier, other learning mechanisms became increasingly important, particularly in-house R&D efforts, to develop new products and to assimilate advanced foreign technologies. By 1988 total private industrial R&D reached around 80 per cent of national R&D spending, amounting to around US$2.5 billion per annum, a three-fold increase over 1981. Around 36 per cent of R&D took place in electronics and electrical technology (*Korea Business World* December 1989 p. 61). R&D investments, motivated by the desire to reduce dependency on Japanese firms, enabled the *chaebol* to develop some new products and to reduce their dependence on OEM and licensing in some areas. The growing complexity of electronics meant that more R&D was needed to enable latecomer firms to absorb technology from overseas. Samsung alone spent more than US$0.5 billion per annum on R&D in the late 1980s.

In-house R&D enabled some firms to negotiate strategic partnerships on a more equal footing with overseas leaders. Later strategic partnerships differed fundamentally from the joint ventures of the early 1970s, where the latecomer firms were junior partners who received technology and training in return for cheap labour. Under strategic partnerships, ROK firms bring technological and other assets to the table to bargain for leading edge technologies, access to markets or both.

Table 4.4   Investments in high-technology US firms by South Korean companies (1986 to 1988)

Year	Company	US Firm	Technology Acquired
1986	Samsung	Micron Technology	US$5m investment, plus license to make 64K RAMs and EPROMs
1988	Samsung	Micro Five Corp	Equity investment for computer technology
1988	Samsung	Comport	Investment for hard disk technology
1986	Daewoo	Zymos	Majority holding for US$13.4m wafer fabrication technology
1986	Daewoo	Cordata Tech.	IBM-compatible PCs, manufacturing and marketing (takeover)
1986	Goldstar	Fonetek	Radio communications, company takeover

Source: Derived from Bloom (1989 p. 28–9).

In the late 1980s licensing gave way to strategic partnerships as South Korean firms grew in competence. In consumer electronics Samsung began joint developments with Tenking (Japan), Thomson (France), TRD (Japan), JVC (Japan) and FROG (Germany). Technologies included VCRs, camcorders and colour TVs. In telecommunications Samsung began joint technology developments with Italtel (Italy), Seiko (Japan) and EMI (US), covering advanced fax machines, office equipment and transmission technology (Koh 1992 pp. 40–41). In 1994 Samsung agreed with NEC of Japan to swap research data on the 256 megabit DRAM, a technology well in advance of the mainstream, to share costs and risks (*Business Week* 14 March 1994 p. 34). Also, as noted earlier, in 1993 Samsung formed an eight year alliance with Toshiba of Japan working together on a new industry standard for flash memories, a technology invented by Toshiba. Samsung had already carried out research into flash memories within its chip division. Its world leadership in conventional DRAMs also made it an attractive partner to Toshiba.

Acquiring Overseas High-Technology Companies

Latecomer firms began investing heavily in overseas high-technology companies during the 1980s (see Table 4.4). Not all of these ventures were successful. Assimilating technology by acquisition was difficult as the Samsung case (Section 4.11) indicates. Some firms set up laboratories abroad within reach of sources of foreign engineers. In 1985, for example, Samsung set up a chip research facility in Silicon Valley, which hired around 300 engineers. Goldstar set up a research facility in Japan in 1981 and in the US (United Microtek) in 1985.

South Korean companies also gained technology through overseas education and training within foreign TNCs. The Samsung Advanced Institute of Technology (SAIT), established in 1987, recruited top level overseas South Korean nationals to manage the new institute and lead R&D projects. SAIT's first president was an ex-employee of a leading US firm (interview with SAIT, 1993). In June 1992, there were at least 60 foreign engineers working with researchers and engineers at Samsung, mostly on a full-time basis (Koh 1992 p. 45). Russian scientists helped develop the green laser technology needed for the prototype of Samsung's digital-videodisk recorder (D-VDR), announced in 1993. Samsung also paid a Russian institute US$320,000 for a new material (MMA) for making plastic optical glass and construction materials (*Korea Times* Tuesday March 16 1993 p. 8).

4.9 SIZE AND SCOPE OF THE *CHAEBOL*

Before examining corporate learning in detail, this section touches on the size and scope of the leading South Korean companies. In the early 1990s three large firms dominated the electronics industry (Samsung Electronics, Goldstar Co. Ltd and Daewoo Electronics), supplying around 44 per cent of total production (calculated from figures presented below).

Goldstar began in 1958 as a 100 per cent locally owned producer of radios for the domestic market. It first exported to a mail order firm in New York in 1962 (Bloom 1991 p. 13). By 1992 it sold products in more than 120 countries and operated nine marketing subsidiaries, ten manufacturing subsidiaries, various joint ventures and five R&D laboratories (*Annual Report* 1993).

Samsung Electronics started by making simple goods in 1969 under a joint venture with Sanyo of Japan. Selling more than US$50 billion worldwide (electronics accounted for around 20 per cent of sales) in 1992, the Group announced its intention of becoming a US$200 billion corporation by the year 2002 (*Business Week* 27 January 1992 p. 22). Meeting this startling objective depended crucially on the company's innovative capacity in electronics. Samsung became the ROK's undisputed technology leader in electronics, although its arch-rival, Hyundai, was the largest *chaebol* overall in 1992. By the early 1990s Samsung had the largest R&D centre in South Korea, manufacturing operations around the world and global leadership in semiconductor DRAM technology.

Daewoo Electronics, founded in 1974, began making small cassettes and car radios for export. In 1983 it acquired the Taehan Electric Wire Company and thereafter diversified into a wide range of electrical and electronics products, establishing overseas manufacturing plants in the UK, France, Mexico and China (*Annual Reports*).

Table 4.5 shows electronics sales for the three companies for 1991 and 1992. Each of the *chaebol* had diversified from simple consumer goods to complex industrial electronics and components. Samsung's sales in 1992 were made up of home electronics (Won 3.3 trillion or US$4.2 billion), industrial electronics (including computers and telecommunications) (Won 1.4 trillion or US$1.8 billion) and semiconductors (Won 1.3 trillion, US$1.7 billion). Total domestic sales were US$6.3 billion in 1992 or 42.6 per cent, compared with exports of US$8.5 billion or 57.4 per cent of sales, illustrating the growing importance of the local market.

In 1990 Samsung spent US$650 million on R&D (a 30 per cent increase on 1989). Although a large amount by most international standards, this expenditure was smaller than Japanese leaders such as Matsushita and Sony who spent US$2.3 billion and US$1.1 billion respectively on R&D (*Asian Business* October 1990 pp. 28–9).

Table 4.5 indicates the small size of South Korea's electronics firms compared with Japanese companies. Although large, Samsung's electronics operations were only one quarter of the size of, say, Sony of Japan. However, company sales were larger than most European and US firms except for IBM (which sold more than US$60 billion in 1992). During the world economic recession of 1992 South Korean firms maintained profitability (albeit low) according to official figures, whereas some Japanese, US and European firms sustained heavy losses. The market recession was less severe in South Korea's low-end areas, while robust local sales helped companies to weather the recession.

Other fast-growing competitors include Hyundai, Anam and Ssangyong. Hyundai Electronics Industries span off from the Hyundai Group in 1983 and by 1993 employed around 12,000 workers, boasted assets of over US$2 billion and in 1992 sold more than US$1.4 billion. The Hyundai Group is a massive conglomerate with sales of US$53.7 billion in 1992, even greater than Samsung's. Prior to 1983, most of its operations were in construction, automotive and shipbuilding, although it had some electronics experience in

Table 4.5   Sales of electronics by South Korean and Japanese conglomerates: 1991 and 1992 (trillion Won, US$ billions)

	1991		1992		1992 net profits	
	Trillion Won	Billion US$	Trillion Won	Billion US$	Trillion Won	Million US$
South Korean firms						
Samsung Electronics	6.0	7.9	6.1	7.8	0.72	92
Goldstar	3.8	5.0	3.8	4.8	0.27	34
Daewoo	1.6	2.1	1.7	2.2	0.17	22
Japanese firms						
Fujitsu	–	29.0	–	30.0	–	0
NEC	–	31.5	–	29.0	–	–100
Sony	–	32.0	–	32.5	–	210
Toshiba	–	39.0	–	39.0	–	200

Sources: South Korean data calculated from *Korea Economic Weekly* 15 March 1993 p. 8; exchange rates for 1991 (as of end of period) US$1 = Won 761; 1992 US$1 = Won 785 (*Korea Business World* February 1993 pp. 52–3). Japanese firms' data from *Fortune* 22 March 1993 pp. 18–19.

its heavy electrical machinery division (e.g. circuit breakers and transformers) (interview and Corporate Profile, 1993).

Anam, which began by packaging simple chips in 1968, became the world's largest chip packaging company, with sales of nearly US$2 billion in 1993. Ssangyong Computer Systems Corporation began in 1981 as a specialist software developer. In 1988 it began making hardware and by 1993 had introduced its own-brand 32-bit microcomputers, page printers and telecommunications systems.

Except for Anam, each of these firms began as offshoots of existing *chaebol*. As a fast growing export market, and with encouragement from government, electronics was an attractive export opportunity for the family owners of the *chaebol*. Within each of the firms technological advance accompanied growth and diversification. By the late 1980s, each operated large R&D facilities and each made sophisticated consumer electronics, industrial goods and components. Most had some own-brand goods in areas such as colour TVs and monitors, camcorders, videocassette recorders, microwave ovens, compact disk players, 16 bit and 32 bit microcomputers, telecommunications exchanges, and 4, 16 and 64 megabit DRAMs.

4.10   LEARNING TO INNOVATE: ANAM INDUSTRIAL

Anam Industrial is almost unknown in the West, despite being the largest chip packaging company in the world. The company carries out sub-contract semiconductor assembly and testing for large, mostly US, firms. Packaging, one of the many process steps involved in chip production, involves encapsulating a tiny integrated circuit with a protective plastic or ceramic coating. Today, it is a highly complex activity with a variety of approaches. Packages include plastic dual-in-line (P-DIP) with 28 to 84 leads (or pins). In 1990 so-called quad, flat packages with 64 to 208 leads were introduced as mainstream. By 1993 Anam assembled around 3,000 types of semiconductors including many leading-edge products. Anam's history is an illustrative example of South Korean latecomer catch up.

Origin and Start up

Anam began as a bicycle producer in 1956. In 1968 the company became the first ROK firm to enter the chip packaging business. The Chairman (Mr Kim) made the decision to diversify after seeing the operations of US chip firms in South Korea. After unsuccessfully trying to sell low-cost packaging services directly to foreign TNCs in the domestic market, in 1969 he instructed his son (then studying economics at MIT) to begin selling directly to parent firms in the US under a new marketing company called Amkor (later changed to Anam/Amkor).

Anam's potential advantage over TNC subsidiaries lay in its ability to use the local highly productive unskilled and semi-skilled workforce to maximum effect. Another advantage over local TNCs was its intimate knowledge of domestic business practices and management.

Milestones and Achievements

In 1970, Anam's first full year of operation, it exported around US$210,000 worth of semiconductors. Exports grew rapidly reaching US$180 million in 1980, US$500 million in 1984, US$1.2 billion in 1990, and around US$1.8 billion in 1992. By 1991, accumulated exports totalled about US$10 billion, mostly through Amkor in the US. By 1993 Anam controlled around 40 per cent of the world sub-contract chip assembly market. In addition to becoming the first South Korean company to produce colour TVs, through a licensing arrangement with Matsushita of Japan in 1973, during the 1970s and 1980s Anam added audio equipment, watches, electric switches and precision instruments to its portfolio. In 1987 it began semiconductor design activities.

In 1993, with sales in the region of US$2 billion, Anam employed around 6,000 staff, of which 4,800 were based in South Korea. Some production had been relocated abroad, as wages and land prices rose at home. In 1989 it purchased a packaging plant in the Philippines from the US firm AMD in order to assemble relatively simple, mature products transferred from the parent plant.

Of Anam/Amkor's marketing outlets in the US, Europe, Japan and Hong Kong, the Santa Clara branch office was the largest. The offices gathered technical information on future packaging needs and some worked jointly with customers. In 1993 Anam's worldwide customers, mostly American, numbered more than 200.

Phase 1: learning the art of assembly

According to Anam's senior engineers, the company's technological development could be divided into four main phases.[14] The first phase, which lasted from around 1968 to 1980, could be called learning the art of assembly. It involved a long slow period of absorbing and mastering the relatively simple techniques of assembly.

Initially, the company imported wafers from TI and RCA in the US, packaged them in the ROK and then re-exported them back to the US. The first products were simple transistors and discrete devices, such as light-emitting diodes. By 1978, the company assembled and tested small-scale integrated circuits.

During phase 1, very little technological know-how was required. Major US clients provided assembly machinery, engineering back-up, detailed specifications and materials. Often, the early assembly equipment was depreciated machinery shipped on boats from America. The clients, who despatched engineers once or twice a year to help with the setting up and running of operations, provided detailed specifications, as with the OEM system. Anam provided efficient labour-intensive assembly services which required little advanced engineering and virtually no on-site development work.

Phase 2: learning process engineering skills

During the second phase, which ran from around 1980 to 1985, Anam learned the engineering process skills needed for more complex products. US clients began to assist with in-house process work around 1980 to ensure standards of quality, productivity and delivery. During the subsequent world chip boom many US companies were unable to meet packaging demands, so they placed larger orders with Anam. At the same time, integrated circuit assembly was becoming more automated and more complex. To gain the skills to meet the growing new demands, Anam invested heavily in engineering training and worked jointly with several of its largest customers.

These initial forays into process engineering culminated in 1984 when Anam set up the Engineering R&D Department (ERD). In contrast with white collar R&D, the ERD's main function was to organize engineering support for manufacturing within the factory. Again, US clients assisted. An engineering expert from TI advised on the operational functions and objectives of the department to maximize its contribution to productivity and quality.

The ERD's objectives were to: (a) provide engineering for new package designs, which involved working with customers on lead frames and substrates; (b) improve, maintain and adapt assembly equipment, which required joint working with suppliers on wafer mount and handling technology; and (c) install, use and modify precision machinery (e.g. the equipment used for producing moulds and dies). ERD was mostly concerned with directing engineering effort to improve and maintain Anam's core manufacturing processes.

Phase 3: the switch to locally initiated learning

The third phase described by Anam's engineers, which occurred between 1985 to 1988, involved a subtle but important switch from customer-pull learning to supplier-push learning. Phase three marked a shift from simple learning to innovation centred on incremental improvements to production processes and products (i.e. semiconductor packages).

Responding to rising technical complexity, Anam took the initiative by offering its own specifications to customers. Where previously the buyers had provided all the technical specifications, many simply outlined their general requirements expecting Anam to produce the detailed design work, alone or jointly. This shift occurred in most mainstream products, but not in the most advanced and more complex packages. Anam's progress in phase three was analogous to the move from OEM to ODM mentioned earlier. By the late 1980s, like many latecomer firms, Anam had internalized both incremental process capabilities and significant design skills.

Customers ceased to be the primary source of engineering competence, as Anam's own engineers took over these tasks. More engineers were hired and some were promoted to senior positions. The ERD purchased new production equipment and installed and adapted it as needed. Company engineers became competent in modifying and operating a wide array of complex packaging equipment. Productivity gains were made through many minor improvements to equipment.

The triggering mechanisms which forced the pace of learning in phase three included the growth in numbers of customers, the diversity of their needs and the importance of ever higher levels of quality. To maximize growth, Anam needed to internalize and extend its engineering operations.

The firm believed that continued reliance on clients would have slowed export growth through delays, higher operating costs, and by dulling, if not damaging, the reputation of the firm. Investments in engineering increased Anam's market confidence and improved its image as a major international supplier of high-quality services, while incremental innovations increased efficiency and profitability.

Phase 4: towards product innovation capabilities

Phase 4 began around 1988 and was still in progress at the time of the research. This phase saw Anam consolidating and improving its product innovation capabilities. The ERD increased its development work on new packages and processes with several leading chip makers, including IBM, TI and Motorola.

Anam's strategy was to exploit its emerging position as a large, world specialist in packaging, selling itself as an international leader in its one chosen area of the semiconductor process chain. Working more on leading-edge assembly techniques, Anam required more analytical and modelling skills for advanced semiconductor packages. Again US firms worked jointly on specific projects to ensure quality and delivery. By offering more new packages, Anam further extended its own product innovation capabilities. The motivating force was to maximize export sales by capturing future market demand.

By 1993 Anam's capabilities included volume manufacture of 256 pin count integrated circuits, covering complex products such as ASICs (CMOS and ECL) and reduced instruction-set chips (RISC). The company operated the latest surface mount packaging under one manufacturing control system. It used statistical process control techniques, computer modelling of production processes, data analysis for product reliability and computer controlled failure analysis. It had also started designing chips for customers.

The ERD was organized as a profit cost centre, responsible for its own revenue targets. Able to offer clients customized design services for new packages of up to 504 pin counts (the leading edge of the technology), the ERD worked with several equipment suppliers from Japan and the US, modifying systems to ensure reliability and to meet new specifications. Most of Anam's mature production (e.g. 14, 16 and 20 pin counts) had been transferred abroad to Anam Philippines Inc.

Anam employed more than 200 engineers in 1993, of which 70 were centralized in the ERD laboratory. The laboratory was equipped with sophisticated CAD facilities for modelling and designing new packages. The director of the ERD had a Ph.D. in physics from an American university where he had worked as a professor. Although the engineering contingent was small by international standards, it provided Anam with a cogent, customer-oriented, innovation capability.

Interpretation: the Nature of Latecomer Innovation

In some respects Anam is typical of the East Asian latecomer advance from minor process innovation to incremental product innovation along a fairly well-established trajectory (in this case, chip packaging). By the early 1990s, Anam, like other latecomers, was able to extend the boundaries of its chosen technological niche. Over the years it acquired the skill to focus its technology investments precisely on the needs of its export market customers.

Anam's progress is not only an example of learning and catch up, it is also a case of innovation capacity building as defined in Chapter 3. Innovation occurred in at least three important senses. First minor *process* innovations, such as capital equipment modifications, enabled Anam to gain productivity and quality improvements. This did not occur overnight, nor as the result of a single strategic decision. As a natural extension of Anam's long experience in labour-intensive operations, this progress enabled the company to gradually assimilate foreign skills and techniques, while earning revenues from exports.

Second, Anam began to make minor *product* innovations. Like many other East Asian firms, Anam learned by relating products to processes, by reverse engineering and by working closely on product specifications with its clients.

Third, innovation characterized the company's *organizational* structure and orientation. Becoming the large-scale supplier of packaging only services to chip makers was an important innovation in its own right. Recognizing and creating a new niche export market led to Anam's strategic positioning as the world's largest chip packaging firm. Many other examples of new organizational approaches were pioneered by East Asian firms. In Taiwan, for example, the company TSMC became the first firm in the world to offer fabrication only services in semiconductors. Other organizational innovations include OEM and ODM, new systems exploited by latecomers to their advantage.

One feature of Anam, typical of East Asian latecomers, was the importance of *foreign channels* of technology transfer. In each phase foreign linkages were exploited to Anam's advantages and to the benefit of its clients. The company's most recent approach to acquiring know-how was to forge joint alliances with international leaders. In Anam's case, foreign connections occurred under its sub-contracting operations. In other East Asian cases it occurred under OEM, ODM or in joint ventures. By exploiting foreign connections Anam followed a path of export-driven technology accumulation, allowing users to provide a finely tuned focusing device for technological choice, learning and innovation. Even in 1993 the company's R&D department was geared almost exclusively to the needs of export clients.

Another typical East Asian feature was Anam's continuing latecomer orientation. Anam rose from bicycle maker to the world's largest semiconductor packager in two decades. In its own area it had become a well-respected, competent firm, having overcome many of its latecomer disadvantages. However, in its structure, strategy and orientation, it remained a latecomer company, not yet a follower or leader. Depending on foreign firms for technology and market outlets, Anam's brand image was conspicuously weak for a US$2 billion corporation. Some of the companies it relied upon for technology were its natural competitors, themselves in the business of chip packaging (e.g. IBM). Lacking strong R&D resources, Anam was not in a position to compete at the innovation frontier as a leader.

4.11   SAMSUNG: INNOVATION LEADERSHIP IN SEMICONDUCTORS

Samsung, the dominant player in South Korean electronics, provides an example of innovation leadership in some areas and latecomer status in others. Unlike Anam, Samsung achieved a well-recognized brand image and introduced major product innovations (in semiconductors). Sections 4.11 to 4.14 contrast Samsung's learning strategies in chips, consumer electronics and telecommunications. Each case proceeded with trial-and-error exploitation of foreign linkages. New forms of learning were devised to allow Samsung to compete at, rather than behind, the innovation frontier in the early 1990s.[15] Despite its progression from sub-contractor to major world competitor, the company confronted latecomer disadvantages in key operating areas.

Corporate Profile

Samsung began in 1938 as a trading company dealing in fruit and dried fish. During the 1950s it diversified into sugar, wool, textiles and other consumer products, responding to the government's import-substitution policy. In the mid-1950s the firm began a fire and marine insurance business. Later in the export promotion period of the 1960s and 1970s, Samsung moved into media (1963), property and retailing (1963), insurance (1963), paper (1965), consumer electronics (1969), construction (1977), aerospace (1977) and semiconductors and telecommunications (1977) (Koh 1992 p. 22).

During the 1960s Samsung was awarded the government's coveted prize of the most successful export company in the ROK. Since 1977, Group revenues rose from US$1.3 billion to around US$24 billion in 1987, to more than US$50 billion in 1992 (Kraar 1993 p. 26). The Samsung Group exported around US$10 billion in 1992, roughly 13 per cent of South Korea's total international sales. The company ranked eighteenth in the *Fortune*'s Global

500 and planned to quadruple sales to US$200 billion by 2002. Table 4.6 shows the scope of Samsung's group activities.

In 1992 the Samsung Group consisted of 25 companies. Overall assets were in the region of Won 18.7 trillion (roughly US$23.5 billion). Although a wide-ranging conglomerate the company only ranked as a major international player in the areas of electronics. Other activities such as textiles, heavy industry, aerospace, insurance and trading were less competitive and confined mainly to the domestic market. According to some reports, the newly formed Samsung General Chemicals has yet to show a return on investment (Paisley 1993 p. 65).

Table 4.6   Samsung Group activities 1993

Electronics	21.7 %
Financial and information services	58.4 %
Engineering	9.9 %
Consumer products	6.4 %
Social services	2.5 %
Chemicals	1.1 %

Source: Samsung, cited in Paisley (1993) p. 68.

Throughout the 1980s and into the 1990s the flagship of Samsung was electronics. In 1992 total electronics sales reached around US$8 billion, of which nearly 25 per cent was in semiconductors. In 1993 Samsung became the world's largest producer of metal oxide semiconductors (MOS), selling around US$2.5 billion worldwide. Largely as a result of Samsung's leadership, in 1994 semiconductors became the ROK's single largest export, amounting to an estimated US$8.4 billion (*Business Korea* March 1994 p. 24).

Technological Achievements in Electronics

Samsung Electronics began by assembling simple transistor radios and black and white TVs under a joint venture with Sanyo Electric in 1969. By 1992 Samsung was the first company in the world to produce working samples of the latest 64 megabit memory chip. In 1993 it invested US$1 billion in semiconductors, bringing its total investments in chips to around US$3 billion (Kraar 1993 p. 28). In 1992 Samsung joined with Toshiba (the world's leading DRAM maker) as an equal partner in an eight year alliance to develop so-called flash memory chips, a new advanced technology. In DRAMs, Samsung had become a world leader, having overtaken most European and US companies, as well as several Japanese leaders.

The history of Samsung in semiconductors is one of speed, tenacity, trial-and-error strategy and learning to innovate. As in the Anam case, technology was acquired through the systematic exploitation of foreign channels and willing suppliers. The semiconductor development can be divided into three main phases.

Phase 1: initial entry into semiconductors
The first phase, which lasted from 1975 to around 1983, saw initial entry and a long period of assembly-based learning. In 1975 Samsung acquired the only locally owned chip company in South Korea (Korea Semiconductor) which manufactured CMOS (complementary metal oxide semiconductor) chips for watches. Although production was expanded the venture lost money (Archambault 1991 p. 55).

In 1980, unsatisfied with Samsung's image as a low end consumer goods producer, the then chairman (Lee) decided to launch the company into the international chip business. There was a conscious attempt to use semiconductors to upgrade and expand Samsung's overall electronics operations. Lee acknowledged the riskiness of the decision, but he decided to follow the Japanese example of mass producing DRAMs. Lee visited Japan and the US and set up a team to assess the most effective means of market entry.

To acquire DRAM technology, Samsung contracted Micron Technology of the US which had developed a 64 kilobit DRAM suitable for low-end PCs and other volume consumer goods. Micron licensed its technology to Samsung and in return Samsung invested around US$5 million in Micron.

There were very few trained South Korean chip engineers at that time. Samsung transferred young engineers from other parts of its electronics business to learn about the technology from the US firm. Although some senior engineers were recruited from US universities, most were hired from South Korean universities. Some of Micron's engineers were despatched to the ROK to train Samsung's staff under the licensing deal.

Samsung also learned about DRAMs by forming a company in Silicon Valley (Tristar Semiconductor) in 1983, with an initial investment of US$10 million followed by a further US$60 million in 1984. This strategy mirrored the Japanese pattern of investing in Silicon Valley companies to acquire US technology (e.g. Fujitsu's investment in the Amdahl Corporation). Initially, most of Tristar's engineers were American. However, Korean managers complained of a lack of dedication (*International Management* October 1984 p. 78) among American employees. Samsung's solution was to hire US-trained Asian engineers, mostly ethnic Korean and Chinese. They succeeded in producing the 64 kilobit chip and later the 256 kilobit DRAM in-house.

In 1984 Samsung began production of 64 kilobit DRAMs as well as related products such as the 16 kilobit EEPROM (also licensed from Micron)

and the 16 kilobit SRAM. However, these operations sustained heavy losses
(reaching a total of US$100 million per annum according to one interviewee).

Phase 2: process technology catch up
Phase 2 involved Samsung's assimilation of 256 kilobit DRAM technology
(between 1984 to 1986) and the development of the one megabit DRAM
(*circa* 1987 to 1988). During this period Samsung ironed out many of its
process difficulties and then quickly took control of the design content of
mainstream DRAMs. During this five-year period Samsung caught up with
the DRAMs leaders by heavy investments, trial-and-error learning and foreign collaboration.

Between 1984 and 1986 Samsung planned to market the next generation
chip (the 256 kilobit DRAM) through Tristar Semiconductor. Some of Tristar's
engineers were recruited from Mostek, a major US DRAM company, then
facing difficulties. The engineers succeeded in producing the device in-house
with Micron's assistance, and technologically the venture was a success.
Initially, the 256k DRAM failed to generate significant returns because of the
1985 market downturn and the subsequent fall in the price of memory circuits.

In parallel, Samsung acquired the design of an EEPROM chip from another US firm, Excel Microelectronics. Both Excel and Micron were facing
financial difficulties and were keen to trade technology for financial support
(Archambault 1991 pp. 59–60).

Despite the accumulating losses and the uncertainty of the market, Samsung
continued with its strategy of improving the manufacturing process and catching up in product design. In 1987 it joined the race to lead in the next-
generation, one megabit DRAM competition, mostly contested by Japanese
companies. Previous experience had given Samsung the confidence to compete in the one megabit DRAM market with its own in-house technology. In
Silicon Valley, Samsung had already built a prototype production line for one
megabit DRAMs as early as 1985.

Again, the decision was taken by Samsung's chairman, despite the losses
already incurred and the considerable investment risk. He decided this time
to mass produce the devices within South Korea. Most of the capital equipment was imported from the US. According to executives, this venture was
fairly independent, not receiving strategic support from vendors or users,
although senior South Korean engineers were recruited from IBM and Intel
to assist with the development.

Samsung's engineers managed to design a one megabit DRAM suitable for
mass production using CMOS technology (the previous two devices had
been based upon NMOS, an earlier generation process). Production began in
late 1987 and volume shipments started in 1988. By 1988 Samsung had

invested a cumulative US$800 million in semiconductors with little return (Archambault 1991 p. 61). Looking back, this was a remarkable gamble for a latecomer firm.

In 1988 the prices of memory chips rose sharply as a strong market upturn occurred. Samsung's main product, the 256 kilobit DRAM, nearly doubled in price compared with one year earlier. According to engineers, Samsung's one megabit design and production know-how fed back into the 256 kilobit line. Micron, the original technology provider, attempted to licence back Samsung's 256 kilobit product (Samsung interviews, 1993).

The 256 kilobit DRAM sold in large volumes and recouped much of Samsung's investments in semiconductors. Samsung's total DRAM sales grew from just US$12 million in 1984 to around US$400 million in 1989, to around US$600 million in 1990. Major buyers included US firms such as Tandy, Apple, IBM and other computer makers (*Electronic Business* 25 June 1990 p. 34). Samsung's operating profits on total semiconductor sales of US$1.4 billion in 1989 reached around US$165 million. Eventually, sales of the one megabit DRAM more than recovered Samsung's entire investments in chip technology.

Phase 3: from catch up to design leadership
During phase 3 (*circa* 1989 onwards) Samsung gained design leadership in 4, 16 and 64 megabit DRAMs, achieving parity with Japanese and US companies in DRAM technology and broadening the company's technology base. In 4 megabit DRAMs Samsung overtook most European and US firms and were only just behind the most advanced Japanese suppliers.

As before, the 4 megabit chip was championed by the company's chairman. Some senior engineers were hired from the US to work on specific problems. One, Dr Ilbok Lee, had worked at Intel as an engineer. Lee had a Ph.D. from the University of Minnesota and, in 1989, he became president of Samsung Semiconductor. Other engineers were promoted from within the company. Some funding (around 10 per cent) was provided by the government's ETRI in a collaborative DRAM venture with other companies, but as in the case of the one megabit DRAM, the core effort and investment was in-house.[16]

Almost all of the 4 megabit design and development work was carried out within Samsung. In 1989 Samsung began shipping its own 4 megabit DRAMs, just behind the Japanese leaders. Large users such as IBM, HP, Sun Microsystems and NEC purchased the product in huge volumes, providing feedback on reliability and performance. Intel formed an agreement with Samsung to re-sell Samsung's DRAMs in the US at this time.

During this phase, evidence of minor product innovations in Samsung abounds. The company produced a so-called synchronous DRAM (for use in connection with microprocessors) which it licensed to Oki of Japan. In 1989 and 1990 Samsung undertook a patent swap with IBM, an exchange of SRAM technology for NCR's ASIC technology, and a joint venture with HP and Intergraph Corporation for RISC microprocessor development. In an arrangement with Zilog Inc., Samsung acquired a licence to produce microcontrollers for use in consumer electronics such as VCRs. To reduce their dependency on Japanese firms US companies were keen to cooperate with Samsung, not only in DRAMs but also on other products.

Samsung's Silicon Valley operation helped to broaden the company's product range by acquiring new technologies. By 1990 the plant employed around 450 people of which around 100 were engineers (mostly American; only 15 were Koreans). As well as marketing and sales, the engineers carried out product development work, technology licensing, materials and capital equipment purchasing. The operation also tested some advanced new products for Samsung (e.g. programmable logic controllers, ASICs and microprocessors) in the US market.

To compete at the world innovation frontier, Samsung forged new, more equal forms of partnership with technology leaders such as Toshiba, NEC, TI, Oki Electric and Corning. With TI, it began a joint chip manufacturing unit in Portugal. With Corning, it began work on advanced ceramics for integrated circuits. As noted earlier, in 1992 Samsung joined with Toshiba (then the world's leading DRAM maker) to jointly develop so-called flash memory chips (by retaining stored information when the power is switched off this technology promised to replace hard disk drives). Although ahead of Samsung in flash memories (Toshiba held the basic patent), Toshiba needed Samsung to help establish the technology as an industry standard. Due to last for eight years, the alliance testified to the new status of Samsung as a world leader in semiconductors.

Finally, according to Kraar (1993 p. 28), in 1992 Samsung became the first company in the world with working samples of the latest 64 megabit DRAM. He reported that Samsung had invested US$1 billion in semiconductors in 1993 alone and that, in total, the company had spent US$3 billion on chip technology.

Innovation Leadership in Semiconductors

Samsung presents an unusual example of a latecomer achieving world leadership in a key technology area. As in the case of Anam, Samsung oriented its entire chip operation towards the export market and acted decisively to acquire and improve upon foreign technology. The company took around 15 years to catch up and overtake most of the international players. By 1992 Samsung was no longer a latecomer, or follower, but an innovation leader in DRAMs.

Samsung created an important imitation effect in the ROK. By 1989, Samsung accounted for around 70 per cent of South Korean integrated circuit sales. Others following Samsung's leadership (mostly by licensing foreign technology) included Hyundai (19 per cent of sales) and Goldstar (8 per cent of sales). In the early 1990s each of the companies, though still latecomers, had plans to build its own process and design capabilities following the example of Samsung.

Remaining Weaknesses

The chairman's original strategy for semiconductors had intended to upgrade Samsung's overall electronics capability by the infusion of advanced technology, an aim still espoused in the early 1990s (interviews with Samsung, 1992 and 1993). However, there was as yet little concrete evidence of synergy between DRAM developments and Samsung's in-house systems development. The DRAM strategy was essentially an export-led commodity venture. Few devices were used in-house.

To a limited extent there were minor spillovers from DRAMs to other chip areas including SRAMs, logic devices and gate array ASICs. In the ASIC area, Samsung ranked nineteenth in the world in 1993. In general though, the DRAM competence contributed little to Samsung's efforts to develop the design-intensive, customized key components needed for consumer and computer products. Samsung remained dependent on Japanese and US firms for many of these. In addition, neither Samsung nor other local firms were capable of designing and producing the capital goods required for chip manufacture. Samsung still lacked the deep technological roots and the synergies which benefited leading Japanese innovators in semiconductors.

Semiconductor Interpretation

The phases of Samsung's development mirrored that of Anam's: a long assembly-based learning period, followed by manufacturing process developments and, finally, evidence of product innovation capability. Samsung pursued a strategy of export-led technology development as proposed in Chapter 3 by using demanding US buyers to pull technology forward through their insistence on high quality, low cost and fast delivery. Had Samsung oriented its production towards local market needs, it is highly unlikely that it could have achieved its transition from latecomer to leader.

More than any other example, the case of semiconductors graphically demonstrates the effort, risk and expense faced by latecomers who attempt to become leaders in electronics. Samsung managed this in DRAMs, but at great corporate risk. Few other latecomers could afford the systematic, long-

term investments in foreign technology sources. However, by in-house efforts, training and its ability to learn from failure, Samsung reached the
innovation frontier in DRAMs and began to form strategic partnerships with
foreign companies to extend the breadth of its leadership to other product
areas.

4.12   SAMSUNG: FOLLOWING IN CONSUMER ELECTRONICS

In consumer goods, although enjoying considerable success, by the early
1990s Samsung still suffered from its latecomer status, relying on OEM/
ODM channels for both distribution and core components. As noted above,
Samsung's consumer electronics business began in 1969 with a joint venture
with Sanyo Electric following consultation with several Japanese firms (including the chairman of Sanyo Electric).

To acquire production skills, the company sent 106 employees to Sanyo
and NEC for training in assembly methods for radios, TV sets and a range of
components (Koh 1992 p. 23). In 1970 under a joint investment with NEC,
Samsung-NEC was formed (the name was changed to Samsung Electron
Devices in 1974).

Up until the mid-1980s, Samsung learned production techniques under
OEM and licensing deals. In 1981 Toshiba licensed microwave oven technology to Samsung. In 1982 Philips supplied colour TV technology. VCR technology was licensed from JVC and Sony in 1983. By the late 1980s, Samsung
had acquired sufficient capabilities to begin joint developments in VCRs with
Tenking of Japan (1989), in camcorders with TRD of Japan (1990) and in
colour TVs (with a German company in 1990) (Koh 1992 p. 22). By 1992
Samsung's main exports were videocassette recorders and microwave ovens
(treated as an electronics product by Samsung). One in five microwave ovens
sold in the US were made by Samsung in 1992, mostly under licensing
arrangements with GTE. Samsung also developed its own brand, low-cost
fax machines which sold in large volumes in the UK and elsewhere.

Although a large proportion of Samsung electronics was still conducted
under OEM, during interviews in 1993, production engineers claimed that
much of the design content of many products was carried out within Samsung
under ODM, rather than basic OEM (see Section 4.8 above). This shift to
ODM occurred in the latter half of the 1980s, signifying that design capabilities had been developed, acceptable to the standards of foreign brand leaders.
In mainstream goods most of the detailed design specifications, design-proc-
ess interfacing and production tooling were under the control of Samsung's
engineers.

However, in the most advanced consumer goods Samsung could boast few if any major successes. The company had made large efforts to reach the product innovation frontier, but had no in-house innovation equivalent to say a camcorder, a Walkman or a colour printer.

The company's strategy was to close the innovation gap through R&D expenditures and deeper foreign technology transfers. For example, it hired 12 Russian scientists to help develop the new digital-video disk recorder (D-VDR) mentioned above. This project used Russian green laser expertise to compress digital data onto a disk space in order to show a feature-length film on a 5.25 inch disk. The system was in experimental use by 1992. Sixty researchers had worked full time on the project which had cost an estimated US$60 million over three years. Another product developed in 1992 was a low-cost (around US$600 per unit) colour video printer capable of printing images from a TV faster than Japanese models. This was developed jointly with Kodak.

The other part of the strategy was to harness the company's advanced semiconductor technology to develop customized chips for systems. However, as argued in Section 4.11, this policy had failed to deliver any observable results. Most of the chip effort had been directed to the commodity export market.

Samsung lagged behind the Japanese leaders in consumer electronics in the early 1990s, still dependent for core components and lacking major, internationally successful product innovations. Japanese firms supplied many of the gas plasma displays (used in some advanced TVs), liquid crystal displays (used in laptops and other PCs), charged coupled devices (used in camcorders) and other customized chips.

To sum up, the consumer case shows that, despite considerable advances, Samsung remained a latecomer, relying on OEM/ODM and licensing for leading-edge products. In mainstream and mature products it had achieved parity, introducing its own incremental product innovations, but it had yet to close the technology gap with the leaders. The company strategy of investing heavily in R&D for new systems and key components had yet to come to fruition, as it had in DRAMs. Needless to say, other South Korean latecomers lagged behind Samsung in consumer electronics.

4.13 SAMSUNG: FAST LEARNING IN TELECOMMUNICATIONS

During the 1980s, Samsung sold large quantities of telecommunications equipment into the domestic market. However, it had little success in export markets and lagged behind European, Japanese and US leaders in high-end equipment.

Entry began in 1977 when the government urged Samsung to respond to the growing domestic demand, especially for public switching gear. Prior to this Samsung had little experience in exchange technology. In 1974 it had begun a joint venture with GTE for private automatic branch exchanges (PABX) which started production in 1975. In 1977 Samsung–GTE was formed, largely for the transfer of PABX technology.

Samsung then acquired the government-owned Korean Telecom Co (KTC) in 1980 to begin work on switching technology. KTC had already established a wide-ranging technology licensing link with ITT/BTM (Bell Telephone Manufacturing) in Belgium. Samsung also began licensing the Metaconta space division switching system (developed prior to fully digital, time-division exchanges) from Alcatel of France to supply the local market.

In order to learn how to make public exchanges Samsung despatched around a hundred engineers to ITT/BTM Belgium for training, while 30 or so Belgian engineers were despatched to help set up the switch manufacturing plant at Gumi in South Korea. Manufacture began in 1980, and by 1993 around four to five million lines had been installed locally. In 1982 Samsung renegotiated with ITT/BTM for the transfer of the fully digital (TDS) System 12, the Korea Telecom Authority (KTA) also pushed Samsung to secure further technology transfer from ITT/BTM.

In parallel with these efforts, in 1980, ETRI promoted a joint venture for an indigenous South Korean switching system (the TDX)[17] and, in 1982, 25 engineers were sent to Belgium for training in the manufacture of time-division mutliplexing exchanges. Goldstar, another collaborator in ETRI, sent their engineers to AT&T for training in the US. Both Samsung and Goldstar used their long-standing relationships with foreign partners to acquire telecommunications technology and to feed this into the domestic TDX system.

The main technology supplier for the TDX system under ETRI was L.M. Ericsson of Sweden. Production of the Ericsson AXE digital exchange had began under licence in 1983. Later, Ericsson was involved in a joint venture called OPC (Oriental Precision Company) with a number of South Korean firms. Ericsson helped develop and produce the localized TDX-1 public exchange (10,100 to 20,000 line capacity) and transferred know-how for the more advanced TDX-10 (100,000 line) exchange.

In the early stages of learning to manufacture it was difficult, but not impossible, for Samsung to hire production engineers. When Samsung began in telecommunications it had no specialist engineers in-house. Prior to 1977, very few South Korean engineers had any experience in telecommunications switching. In 1978 young electronics engineers were recruited from Samsung Electronics and other parts of the Samsung Group to begin learning the technology, as in the case of semiconductors. Samsung also poached some engineers from Goldstar, which had operated a joint venture with Siemens for the mechanical Strowger systems. The company managed to recruit several engineers from OPC as well. Formal in-house technology training began in 1978, but the core learning took place in Belgium with the team made up mostly of young engineers.

Although many studies stress the importance of government intervention in South Korean telecommunications development under ETRI, the chief source of technology was foreign firms. Samsung and the other latecomers bargained with foreign companies for training and technical assistance inside and outside of South Korea. As the instrument of the government, ETRI oversaw some of the software development and organized field trials. Key components (e.g. microprocessor chips) had to be imported from ITT and AT&T and assimilated by firms such as Samsung.

Studies of South Korean technological progress, especially in telecommunications, often overlook the strategies, skills and efforts made by local firms in setting up and exploiting foreign channels of technology to their advantage.[18] By viewing company behaviour and technology acquisition as merely a response to government policy, such studies underestimate the tenacity, strategy and ability of South Korea's latecomer firms.

Comparing Samsung's Technological Learning Paths

The semiconductor, consumer and telecommunications cases testify to Samsung's strategies, strengths and weaknesses in electronics. They show how Samsung progressed from simple OEM through to licensing, ODM and joint developments of technology with market leaders. Samsung's in-house engineering and R&D efforts helped it acquire, adapt and improve upon foreign technology. Learning gradually from competitors, Samsung took a long-term strategic approach to technological acquisition.

The cases show that, with the exception of semiconductors, the company remained a latecomer, dependent on its foreign suppliers for complex software, key components and, frequently, export market channels. In telecommunications Samsung relied on outside suppliers for core software skills and for customized chips, although the gap with the market leaders had narrowed through time. In consumer goods too, the company was behoven to its natural competitors on the world stage. In semiconductors, weaknesses remained in customized systems chips and capital goods.

4.14 INNOVATION MANAGEMENT CHALLENGES

While this chapter is mainly concerned with corporate strategy, the question of how innovation management takes place within latecomer firms is another poorly understood and seldom analysed topic. Although the subject raises important questions of history, culture and best practice, here it is only possible to touch on some of the key management problems and challenges facing the *chaebol* as they proceed towards the innovation frontier.

As latecomers, South Korean companies modelled their organizations and strategies on those of Japanese firms, relying on large-scale manufacturing and conglomerate structures. Indeed, the Chinese characters for *chaebol* (business group) and Japan's pre-war *zaibatsu* are the same. The *chaebol*'s process-led strategy delivered rapid technological learning, leading to innovation and economic growth. The approach helped raise South Korea from poverty to full employment and rising wages. Family entrepreneurs and managers took risks, built up national technology competences and achieved export market growth. South Korea's managers proved to be notoriously hard working, highly motivated and disciplined. Strong leadership gave direction and focus to industrial development and established the *chaebol* as the main engine for industrial growth in the ROK.

However, the strategy brought with it structural difficulties. In contrast with the Japanese *keiretsu* (mutually supporting associations of firms), by the early 1990s the *chaebol* had failed to develop the technological roots of the Japanese companies. The *chaebol* lacked the supply chains of specialized SMEs found in Japan. While the manufacturing-led strategy of extreme vertical integration helped overcome entry barriers, it left Samsung, Anam and others weak in product innovation capacity, advanced R&D, science, key components and new materials.

The particular historical path of the latecomers led on to the innovation management problems facing the *chaebol* in the early 1990s. Huge size, bureaucratic structures, unwieldy conglomeration and rigid hierarchies were commonly recognized company problems in the ROK.[19] While early strategies and structures had enabled Samsung and others to mobilize large resources and to cross-subsidize risky operations, the old approach became a liability, preventing firms from competing at the innovation frontier. Through the 1990s, the *chaebol* will have to make major restructuring changes if they wish to overcome their latecomer legacy.

According to managers, the scale-intensive formula restricted corporate flexibility and slowed responses to market changes. By contrast, Taiwanese firms (see Chapter 5) were more able to move quickly into new markets and raise their product innovation capacity. South Korean firms were relatively slow to relocate investment to low-cost areas of East Asia, making some of their goods uncompetitive. The cross-subsidizing practices masked cost difficulties, while bureaucracy stifled the innovation capacity of some young South Korean engineers and researchers.

As South Korean firms approached the innovation frontier some have attempted to become more flexible and innovative, but challenging the legacy of the past is difficult. Rigid hierarchies, promotion based on age rather than merit, authoritarian management styles and concentrated decision making are deeply rooted problems. Even among the most modern firms there is still a high degree of formalization and centralization in decision making, compared with leaders such as HP or DEC. For example, even small spending decisions are often routed through formal procedures (called *kyul-jae*) which not only introduce delays but are used by layers of managers to exert bureaucratic control and authority.[20]

The strong leadership of owner-managers often reinforces the authoritarian style and slows or prevents the decentralization of power and the diffusion of modern management techniques. In most companies ownership and management have yet to be separated, reflecting the latecomer status of the *chaebol*. Owner families are still the most powerful industrial force in South Korea. One study of 108 ROK business groups showed as many as 8.8 per cent of those in top management are related to the owner (Lee 1989 p. 156).

The old structures and strategies succeeded in the historical task of mass producing standard goods for predictable markets. However, they fall short of dealing with fast-moving, complex high-technology industries and niche markets. Old practices restrict the creative input from young engineers and R&D staff, while centralized communications along vertical hierarchies reinforce old methods. Strict seniority prevents subordinates from feeling free to communicate openly with managers. As a result, cross-functional cooperation is lower in South Korea than in Japan and relations are even more formal than in Japanese firms, let alone US companies (Chung and Lee 1989 p. 174).

Latecomer weaknesses are also evident at the broader corporate level. There is evidence that profitability is relatively poor in South Korean firms, compared with US and Japanese firms and that companies diversified too far into non-core business areas (Chung 1989 p. 6). Many firms still depend heavily on government-controlled banks for finance and on market protection provided by government.

Most of the difficulties mentioned here are well recognized in South Korea. There is evidence of change and a clear recognition of the need to become more flexible among many firms. The leaders, such as Samsung, have attempted to climb up the technology ladder, to decentralize their operations and to introduce performance-based reward systems. Companies such as Daewoo and Goldstar have introduced high-cost foreign consultancy programmes to increase management professionalism. If these efforts are successful the next decade will witness the emergence of more flexible, market responsive and innovative corporations in the ROK.

4.15   CONTINUING LATECOMER ORIENTATION

This chapter has shown how South Korean latecomer firms substantially narrowed the gap in electronics. Through their strategies of closely linking technological learning to export market demand, the *chaebol* created their powerful presence on the international stage. Within large vertically integrated corporations, the needs of demanding export customers were used as a focusing device for technological investments and efforts.

The evidence demonstrates how firms progressed from learning the techniques of simple manufacturing to genuine innovation. Unlike R&D and design-led innovation, typical of world market leaders, South Korean latecomer innovation concentrated on incremental improvements to manufacturing processes. Continuous minor manufacturing innovations led to product design and development. Consistent with Chapter 3's propositions, only in the latter stages, and only selectively, did R&D become a significant weapon in the innovation armoury of the latecomers.

The chapter stressed the importance of organizational innovation and especially links with foreign companies. Beginning with the OEM and subcontracting, firms learned from TNCs and other buyers. By the late 1980s the OEM system had evolved into ODM, indicating new latecomer competences in design. The OEM/ODM channels which allowed systematic technological learning did not simply arise from nowhere: they were pioneered, developed and improved upon by local firms to the advantage of their foreign partners. These strategic, innovative efforts enabled firms to surmount their extreme latecomer disadvantages.

The latecomer origin and orientation of local companies pinpointed their continuing weaknesses. In a small number of areas firms had introduced major new product innovations, but in most fields they were dependent on their natural competitors for key components, capital goods and distribution channels. Weak in R&D and lacking a high-quality brand image abroad, most firms continued to rely on catch-up, imitation-based growth. The *chaebol* had yet to establish deep roots into capital goods, advanced materials and key components, while the weakness of local SMEs compounded the problem.

Nevertheless, it would be wrong to overstress the problems facing South Korea's latecomers.[21] These almost pale into insignificance compared with the difficulties already overcome since the 1960s and 1970s. As local firms approach the technology frontier they have forged strategic partnerships with world leaders to acquire highly advanced technologies. Brand-name goods continue to increase as a proportion of total production, R&D spending has grown, and the *chaebol* have narrowed some of the innovation gap. Because their learning achievements are cumulative and built upon solid foundations, South Korean firms are well positioned to create new market opportunities and to respond to the fast-changing pace of electronics technology.

NOTES

1. The terms South Korea and Republic of Korea (ROK) are used interchangeably in this book.
2. Around 55 interviews were conducted with companies and government organizations during three research visits to South Korea. A small selection of the case studies is used below to highlight South Korean latecomer strategies.
3. This is most evident in the early stages. In the later stages, firms relied far less on government as shown below.
4. For the wide range of specific government policy measures see Lim (1992 p. 5).
5. It should be noted that there are significant differences in structure, finance and organization between the South Korean and Japanese conglomerates (Whitley 1992).
6. As Amsden (1989 p. 14) points out, Park Chung Hee (president from 1961 to 1979) allowed the 'millionaires' who promoted his reforms into the heart of state decision making. Park wanted these tycoons to create large plants to realize economies of scale and to promote national capitalism. According to Park's philosophy the role of government was to oversee the millionaires to prevent any abuse of power.
7. Hobday (1991) provides a detailed historical review. Also see Paek (1992), Kim (1989), Lim (1992) and MOST (1993).
8. Data from the Electronics Industry Association of Korea, cited in Korea Exchange Bank (1980 pp. 172–80).
9. For details see Korea Development Bank (1988 pp. 122–4; original data from Electronics Industries Association of Korea).
10. Korea Exchange Bank (1980 pp. 172–80; original data from The Electronics Industries Association of Korea).
11. Unless otherwise stated the following figures are from *Industry in Korea* (Korea Development Bank 1988 pp. 100–120; original data from Electronics Industries Association of Korea).
12. The following data are calculated from the Statistics of Electronic and Electrical Industries, Production–Export–Import, 1992.4 provided by the Electronics Industries Association of Korea in 1993.
13. Cowley (1991 p. 20) makes this point for Taiwanese firms (see Chapter 5).
14. These details were gathered during a series of interviews with senior staff. The dates are rough approximations and the interpretations are largely that of Anam's staff.
15. Unless otherwise stated the evidence presented is from a series of ten or so interviews conducted between 1991 and 1993 in various branches of Samsung.
16. It is not possible here to assess the role of ETRI which is often cited as being important to the DRAM venture. Suffice to say that during interviews few, if any, of Samsung's engineers mentioned ETRI, unless prompted.
17. ETRI actually developed out of another institute in 1982, the Korean Telecommunications Research Institute (KTRI) initially set up as a branch of the Korean Institute for Science and Technology (KIST).
18. Often this is a matter of emphasis, with authors preferring to focus on the role of government rather than firms (e.g. Amsden 1989; Lim 1992; Kim and Dahlman 1992). In a typical study of telecommunications by Kim et al. (1992) very little attention is payed to firms and much is ascribed to government policy. However, without the entrepreneurialism of South Korean companies, policies could not be successful.
19. All companies interviewed recognized the need for change. Some, such as Goldstar and Samsung, were implementing programmes to alter company philosophies, to flatten corporate pyramids and to increase speed and flexibility in the marketplace.

20. See Lee 1989 (p. 154) for an analysis. Often attempts at change are thwarted by middle
     managers, who see their positions threatened by modern practices.
21. See Ernst and O'Connor (1992) for a very negative assessment of South Korea's future in
     electronics. By concentrating on the general difficulties in competing in modern electron-
     ics, they overlook and underestimate the historical achievements, skills and strategies of
     latecomer firms in the ROK (and elsewhere in East Asia). Focusing mainly on barriers to
     entry, Ernst and O'Connor (1992 p. 267) argue that the NIEs are poorly equipped to meet
     the challenges of the electronics industry, and that they have largely failed to develop
     viable strategic responses to changes in the industry. This is a highly questionable conclu-
     sion given that the NIEs' historical progress, especially in electronics, is almost unparal-
     leled in economic history (Chapter 1). They also argue that NIE firms are facing an
     erosion of competitiveness (1992 p. 270), again an unconvincing conclusion given that
     many East Asian latecomers have overtaken European and American companies in elec-
     tronics and that traditional market leaders (e.g. IBM, Amdahl, NEC and Matsushita) not
     only faced heavy losses and painful restructuring in the early 1990s, but were dependent
     on low-cost East Asian manufactures for their survival. As this book shows, the NIEs
     continue to enjoy large trade surpluses with the West in electronics and they are well
     poised to continue their advance towards higher quality, more complex products. A more
     balanced account of South Korean prospects is provided by Bloom (1989). See also
     Chaponniere and Fouquin (1989) and Gee (1991) for Taiwan. Few of these (and other)
     policy studies examine corporate strategies and histories.

5. Taiwan: small firm innovation clusters

5.1 THE COMPARISON WITH SOUTH KOREA

This chapter examines patterns of technological advance in Taiwan's electronics industry, showing how local firms overcame barriers to entry. In many respects Taiwan provides a stark comparison with South Korea. In Taiwan, industrial development relied on a multitude of small and medium sized enterprises (SMEs).[1] By the early 1990s, even the largest Taiwanese electronics firms were still a fraction of the size of the South Korean *chaebol*.

However, there are also similarities with South Korea. The Taiwanese electronics industry benefited considerably from TNC investments, joint ventures and foreign buyers. TNCs helped to foster the start-up of many of Taiwan's electronics makers as large numbers of local firms supplied them with goods and services, leading to a thriving sub-contracting and OEM system. In contrast with the ROK, in Taiwan FDI continued to play a central part in the electronics sector through the 1980s and into the 1990s.

By 1989 the term ODM (see Chapter 3, Section 4) had begun to be used widely in Taiwan. As in South Korea, Taiwanese latecomers learned to design and manufacture products to be then sold under a foreign buyer's brand name. ODM provided an alternative to own-brand manufacture (OBM) for small firms. Most importantly, ODM signified a new stage in latecomer product innovation, going beyond the processes learned under OEM and subcontracting. Under both OEM and ODM, exports focused domestic learning efforts and helped to pull Taiwan's competitive capabilities forward.

The Taiwanese case shows how hundreds of tiny latecomer firms clustered together behind the electronics frontier to exploit market opportunities, indicating that the large-scale, mass market approach followed by the *chaebol* is not the only route to export success for developing countries.[2] The chapter shows how, in electronics, sewing machines, footwear, bicycles and other fast-growing export industries, small firms made themselves indispensable to foreign buyers and TNCs and forged backward linkages to other industries. A simple backward–forward linkage model is developed (Section 5.11) to illustrate how local companies attracted more foreign buyers, stimulating further domestic entry, and ultimately producing a number of thriving industrial export clusters, operating behind the innovation frontier.[3]

Taiwan's progress from OEM to ODM and, in some cases to OBM (Section 5.8), did not occur without difficulties. Indeed, the latecomer origin and orientation of Taiwanese firms provides deep insights into the economy's technological and market weaknesses. However, the dedication of local companies to export markets suggests that they are well poised to meet the challenges of the future.

5.2   THE HISTORICAL CONTEXT

Taiwan's Traders and Sub-Contractors

The divergent corporate structures of Taiwan and South Korea can be partly explained by each country's starting positions in the 1940s and 1950s, and by how their governments responded to these initial difficulties. Taiwan was a colony of Japan from 1895 to 1945. In contrast with South Korea, a range of Japanese-owned industries, including textiles, cement and petroleum refining, was established in Taiwan.[4] During the period 1912 to 1940 manufacturing grew at a healthy rate of around 6 per cent per annum (Wade 1990 p. 74). The trading sector also developed, as Taiwan was used by Japan as a centre for processing raw materials imported from South East Asia which were then re-exported to Japan.

Under the Japanese, economic and social developments far surpassed those in South Korea. In Taiwan, the mortality rate fell, schools were set up and by 1940 almost 60 per cent of children attended primary school. Agriculturally oriented two-year secondary schools were established in most towns, and many Taiwanese gained experience in the technical professions and management. The native Taiwanese who served under the Japanese were later to become an important source of industrial entrepreneurs. By the mid-1940s Taiwan was already more commercially and industrially advanced than most of the provinces of China (Wade 1990 pp. 74–5). However, it should also be recalled that in 1949 Taiwan was still a very poor country with average annual per capita income below US$100, roughly the same as India's (Vogel 1991 p. 13).

In 1945, after the Japanese defeat, Taiwan reverted back to the Republic of China, or more precisely, the Nationalist Party (or Kuomintang, KMT) which was then at war with the Chinese Communist Party. After the KMT's defeat in 1949, Chiang Kai-Chek and the armies of the Nationalist Party retreated to Taiwan (around 150 kilometres from the Mainland) and took control of the economy, the administration and the former Japanese industries. Around two million soldiers and civilians overwhelmed the island, swelling the existing population of around six million.

In the 1950s Taiwan developed rapidly and by 1955 the country's per capita income was around US$140, more than 70 per cent higher than the South Korean's (Levy 1988 pp. 44–5). The percentage of school leavers above the age of 16 with 12 or more years of education was nearly three times higher in Taiwan. South Korea was only able to catch up and overtake Taiwan's education record in the late 1970s.

Compared with South Korea, during the 1950s and 1960s Taiwan's markets functioned relatively well and small traders proliferated. Sub-contracting relationships were formed swiftly in response to export market opportunities, usually identified by traders. As Levy (1988 pp. 44–5) shows, the highly active traders and sub-contractors enabled firms to begin producing electronics with lower initial investment costs than their ROK counterparts. In South Korea the government helped overcome weaknesses in trading by promoting large-scale general trading companies modelled on the *sogo shosha* (general trading company) of Japan.[5] In Taiwan traders kept pace with industrial expansion and by 1973 there were four times the number of export traders in Taiwan than in South Korea. By a wide margin, market failure was less pronounced in Taiwan than in the ROK.

Early Government Policies

In sectors of low capital intensity such as electronics assembly and textile manufacture, the government intervened less than in large-scale, complex technological fields and intermediate goods. In the latter, successive Taiwanese governments promoted industry through direct intervention and financial support. During the 1950s, up to one half of Taiwan's industrial production took place in state-owned enterprises (Chaponniere and Fouquin 1989 p. 28). In steel, petrochemicals, shipbuilding and automobiles, state enterprise accounted for a large proportion of total investment through the 1960s (Wade 1990 p. 110). As recently as 1987, seven of the 20 largest firms in Taiwan were state-owned.

However, in light industries such as electronics, textiles and plastics the government encouraged the market to work through private enterprise by securing a stable macroeconomic environment, keeping inflation low and by protecting the local market. Until very recently, the electronics industry was protected from foreign competition by import restrictions and other constraints (Chaponniere and Fouquin 1989 pp. 29–30). The government often negotiated the terms of entry of foreign TNCs and devised the export-led industrial policy for latecomer firms to exploit international market opportunities.

In the 1950s and 1960s, the Council for Economic Planning and Development (CEPED) had overall responsibility for the economy while the Ministry of Economic Affairs (MOEA) was responsible for planning industrial development. The MOEA's successive five year plans encouraged private industry, leading to a fall in the degree of government enterprise in national economic activities to around 16 per cent of industrial production in 1985 (Chaponniere and Fouquin, 1989 p. 28).

In order to generate employment and export earnings during the 1960s and 1970s the government offered incentives to both foreign and local electronics companies. During the 1970s and 1980s the state provided technical support to industry through government-owned R&D institutes and universities. The state also invested, selectively, in a small number of companies working in scale-intensive, high-technology upstream sectors such as semiconductors.

The Importance of Government

Although the government is often credited with much of Taiwan's industrial success, few studies systematically examine the costs, benefits, successes and failures of state policies.[6] In electronics, it is likely that direct technological and industrial intervention had little effect during the 1960s and 1970s. Left to the market, the TNCs and hundreds of latecomer firms began the industry by manufacturing simple products. As in the case of the ROK, local firms developed through their outward orientation towards foreign buyers and TNCs, not inwards towards the government or the local market. Many small Taiwanese businesses mistrusted government, feared officialdom and kept their distance from state agencies.[7] It is unlikely that specific programmes led directly to the start-up of many latecomer firms, or affected directly the strategies of large private companies such as Tatung, Sampo and Teco.

Nevertheless, overall industrial policies provided a framework for exporting and incentives to some local firms, as well as to foreign TNCs and buyers. The government orchestrated the industrial and macroeconomic environment of low inflation and high savings, which encouraged latecomer firms. The state also supplied the educational and infrastructural support needed for industrial development. Later on in electronics, the state controlled Industrial Technology Research Institute (ITRI) trained engineers in advanced semiconductors and transferred technology to local firms. ITRI incubated several companies which eventually formed the nucleus of the domestic chip manufacturing industry (see Section 5.10).

While the overall impact and cost of direct intervention is more questionable, the government clearly had some success in assisting parts of the electronics industry to overcome barriers to entry. However, Taiwan's vigour in electronics depended primarily on the strategies and abilities of entrepreneurs, engineers and managers. Without such capabilities, no macroeconomic policies, however well designed, could have produced Taiwan's economic miracle.

5.3 ACHIEVEMENTS IN ELECTRONICS

As in South Korea, Taiwan began by making simple transistor radios and black and white TVs in the 1950s and 1960s. Colour TV production began in the 1970s and during the 1980s firms moved into TV monitors, VCRs and computers. By the early 1990s, the country was a leading producer of colour monitors, PCs, computer boards, terminals and switching power supplies, mostly sold under OEM/ODM arrangements.[8] Taiwan's largest computer makers (e.g. ACER, Wyse and Tatung) produced advanced PCs for US firms, sometimes using their own designs. Smaller firms tended to concentrate on older vintage products under OEM and other sub-contracting arrangements.

During the 1980s, some latecomers grew fairly large. The sales of both Tatung and ACER already exceeded US$1 billion by 1990 while other major firms, including Sampo Corporation, Mitac and United Microelectronic Corporation (UMC), grew rapidly. In 1990 the sales of Taiwan's five leading electronics producers amounted to just under US$3 billion (compared with more than US$15 billion for the three largest South Korean companies), emphasizing Taiwan's fragmented industrial structure.

During the early 1990s, Taiwan's latecomers challenged the market leaders in PCs, workstations, colour monitors and other products. Although foreign-owned firms and joint ventures, once in control of Taiwanese industry, declined as share of total output, in 1989 they still accounted for about 35 per cent of computer related exports (III 1991 p. 37).

Table 5.1   *Exports of electronics: selected years 1970 to 1990 (US$ billions)*

Year	Exports
1970	0.2
1974	1.2
1980	4.1
1985	4.9
1987	10.6
1990	17.2

Sources: Ministry of Finance, MOEA and Taiwan Electric Appliances Association, cited in Gee (1989 p. 3); O'Connor and Wang (1992 p. 41); and Chaponniere and Fouquin (1989 p. 9).

As in South Korea, by the late 1980s electronics was the largest industrial sector, having overtaken textiles in 1986. Table 5.1 shows the growth in exports since 1970 and the take-off after 1985. During the 1980s, electronics contributed between 18 per cent and 20 per cent of total exports, while 70 per cent of production was exported annually. According to the Taiwan Electric Appliances Manufacturers Association (TEAMA), during the period 1985 to 1988 electronics exports grew at an average annual compound rate of 36.8 per cent. By 1987 the sector accounted for around 5.3 per cent of GNP (Chaponniere and Fouquin 1989 p. 9).

The US absorbed the lion's share of Taiwan's exports, followed by Europe, with Japan a distant third. Like South Korea, many key electronics components, materials and capital goods were imported from Japan during the 1980s. The import value of electronics production tended to be heavy, amounting to around 52 per cent in 1980 and 54 per cent in 1988.

As noted, Taiwan's electronics industry consists mainly of small and medium sized firms, defined in Taiwan as firms with less than 300 employees (Dahlman and Sananikone 1990 p. 95). Although some grew fairly large, SMEs dominated exports through the 1980s. In 1985 the average electronics firm employed 24 people, even less than in Hong Kong. In 1992 there were around 3,630 makers of electronics and electrical goods, plus several thousand traders (*TEAMA Directory 1992*). In computer products alone the total number of exporters in 1989 was roughly 3,700, of which around 650 were manufacturers. The rest were mostly trading companies based in Taipei (III 1991 p. 37).

Table 5.2   Electronics and information technology, production values and forecasts (US$ billions)

	Output 1990	Forecast 2000	Average annual growth (%)
Information products[a]	6.9	28.0	15.1
Automation	2.8	10.1	13.5
Consumer electronics	2.3	4.5	7.0
Telecommunications	1.9	8.2	16.0
Semiconductors	1.5	6.0	14.8
Total	15.4	56.8	

Note: [a] Defined as microcomputers, terminals, colour monitors, disk drives, peripherals and other computer related components and products.

Source: Derived from data presented in O'Connor and Wang (1992 p. 42).

Table 5.2 shows Taiwanese electronics production for 1990 and plans for the year 2000. During the 1980s, information (i.e. computer) products overtook consumer goods as the largest sub-sector, growing from around US$80 million in 1980 to US$2.1 billion in 1986 to nearly US$7 billion in 1990, by which time it formed around 3.8 per cent of Taiwan's GNP. Nearly US$5.9 billion (96 per cent) of total computer production in 1990 was exported (40 per cent to North America, 41 per cent to Western Europe and 14 per cent to Asia Pacific). Japan accounted for only 2 per cent of Taiwan's computer exports in 1990 (O'Connor and Wang 1992 p56; Institute for Information Industry, Market Intelligence Center data).

Compared with South Korea, Taiwan graduated further from consumer electronics. While the ROK was more dependent on video equipment and colour TVs in the early 1990s, much of Taiwan's consumer-good and other low-end production had been transferred to China. Conversely, South Korea was ahead in semiconductors, exporting around US$5 billion in 1990 compared with Taiwan's production of around US$1.5 billion.

Differences in product specialization reflect each country's industrial structures and historical experiences in electronics, rather than overall technological leads or lags. In the early 1990s, Taiwan was ahead in computer technology, PCs and integrated circuit design (e.g. ASICs), while South Korea led in very large-scale integrated circuit technology and high-end consumer goods (e.g. camcorders). Taiwan had made more headway in software development for systems use, while the *chaebol*'s scale advantages gave them a lead in semiconductor fabrication technologies. Both countries exported similar quantities of colour monitors. In areas where corporate size and production scale conferred advantage, the *chaebol* tended to be ahead of Taiwanese firms. Where speed, flexibility and design were more important, Taiwan tended to lead.

The two main computer exports from Taiwan in 1990 were PCs (US$1.6 billion) and colour monitors (US$1.3 billion). Other computer peripherals, including keyboards, mice and scanners amounted to US$2.2 billion in 1990. In 1990 Taiwan, ranked seventh among computer goods producers worldwide, boasted around 700 hardware manufacturers (mostly SMEs) and about 300 software and service companies (O'Connor and Wang 1992 pp. 53–4).

Taiwanese firms gained large market shares in certain products (Table 5.3) including around 83 per cent of the world's motherboards (the main circuit board used in PCs). In 1993 Taiwanese motherboards were used in roughly 40 per cent of PCs sold throughout the world (*Electronics* 10 January 1994 p. 11). Similarly Taiwan was the largest producer of PC mice, monitors, image scanners and keyboards. In 1993 and 1994 computer hardware production grew at around 15.5 per cent and 14 per cent respectively to reach an estimated output of US$10 billion, plus software services of around US$1.4 billion.

Hundreds of flexible, fast moving small firms entered the computer industry. By the early 1990s, despite many exits, the industry had earned itself the reputation of being the international arms dealer of the computer trade (*Business Week* 28 June 1993 p. 36). Taiwan's SMEs learned to adjust to new markets and to respond quickly to the OEM, ODM and sub-contracting needs of larger foreign firms. In 1994 computers (including notebooks and desk-top PCs) surpassed peripherals to become the leading export item, a further step forward in the industry's technological advance.

During the 1960s and 1970s consumer goods led Taiwan's electronics industry. As firms gained more sophisticated capabilities and costs rose, consumer output declined from US$4.4 billion in 1987 to US$2.3 billion in 1990. Firms relocated low-end production to China (especially to the nearby Fujian Province), upgrading their own operations. Although the growth in world demand for consumer electronics had slowed, Taiwan continued to export many consumer goods including colour TVs (US$505 million), calculators (US$390 million), VCRs (US$198 million) and electronic clocks (US$193 million) (data for 1989; O'Connor and Wang 1992 p. 68).

The telecommunications industry, which pre-dated consumer goods, started in 1957 when local companies began to produce telephone handsets and switches for domestic use. Exports grew rapidly after the liberalization of the US market in 1983. Low-end consumer-type products such as digital telephones, key systems, fax machines, PABXs and modems were exported in

Table 5.3   Taiwan's leading computer products, 1993

Product item	Production (millions of units)	Market share (% world)
Monitors	17.5	51
Desktop PCs	2.3	8
Notebook PCs	1.3	22
Motherboards	12.3	83
Cathode ray tubes	1.4	24
Graphic cards	7.1	31
Switching power supplies	21.2	30
Image scanners	1	55
LAN cards	3.8	27
Mice	22.1	80
Keyboards	18.8	49

Source: Institute for Information Industry, Market Intelligence Center data, cited in *Electronics* 10 January 1994 p. 11.

bulk to the US and Western Europe. By 1992 Taiwan's share of the world-wide fax machine market was estimated at 50 to 60 per cent and rising (*Computrade International* 15 May 1993 p. 88), again mostly under OEM and ODM.

Around 97 per cent of locally produced public telecommunicatons ex-changes in 1990, as in South Korea, were for domestic use, rather than exports (O'Connor and Wang 1992 p. 43), reflecting the weakness of Taiwan-ese firms in complex hardware systems and advanced software.

5.4   PHASES OF ELECTRONICS DEVELOPMENT

Start-up Phase (1950s and 1960s)

As Wade (1990 pp. 93–4) shows, Taiwan's electronics industry originated in the late 1940s when local job shops began to assemble radios using imported parts from Japan. Other firms which had transferred from Mainland China began to make simple electrical equipment such as transformers, wire and light bulbs. In 1950 the government restricted the import of finished radios in order to protect local manufacturers and provide incentives to component producers.[9] Taiwan's earliest technology licence agreement, for electric watt-hour meters, took place in 1953 between Tatung and a Japanese firm. As with South Korean companies, Tatung sent its engineers to Japan for training.

In the late 1950s large TNCs began searching abroad for cheap labour locations. Taiwan's historical links with Japan led to several investments, while the country's role as a counter to Chinese communism favoured US FDI. The skilled labour force, government incentives and US aid programmes all encouraged US firms to invest in Taiwan at a time when Latin American countries were restricting TNC activities and insisting on local ownership (Wade 1990 pp. 149–50). From Japan, leading foreign investors included Sanyo, Matsushita, Orion, Sony, Sharp and Hitachi. From the US, General Instruments, TI and DEC were early entrants.

After experimenting with import-substitution, between 1958 and 1962 the government introduced a series of incentives to promote exports. The first free trade zone offering cheap labour and tax incentives to foreign firms was opened in 1966 by the port of Haohsiung. Two more followed in Taichung and Nantze. By 1986 the zones had attracted 96 electronics producers who employed some 45,000 people (Chaponniere and Fouquin 1989 p. 29 and p. 41).

Although aggregate FDI flows into Taiwan were small, foreign companies took the lead in electronics through wholly owned subsidiaries and joint ventures (see Section 5.6). In 1954 NCR became the first US company to begin operations in Taiwan (Gee 1989 p. 1). In 1964 General Instruments began making consumer electronics and later transferred production of transistors, diodes and integrated circuits from the US (Chaponniere and Fouquin 1989 pp. 29–31). In the following two years, 24 more American companies began to make simple electronics components and other products for export.

By contrast, Japanese firms tended to supply the local market through joint ventures. By 1963 at least seven formal joint ventures had been agreed between local and Japanese electrical appliance manufacturers (Wade 1990 p. 94), mostly producing transistor radios, black and white TVs and simple components. In 1963 Sanyo formed a joint venture with the Taiwanese importer of its goods to supply the local market and by 1970 the company began exporting. This venture initiated production of white goods, air-conditioners, audio products, TV sets and, later, VCRs (Chaponniere and Fouquin 1989 p. 47). In 1968 Sanyo set up a factory in the Taichung free trade zone which employed around 3,500 people by the mid-1980s. Other Japanese firms soon began exporting after setting up in Taiwan.

The Taiwanese Government's initial motive for encouraging electronics, as in South Korea and the other dragons, was to create employment, earn foreign exchange and exploit what was a fast-growing export opportunity. In the early days there was little strategic thinking about the place of electronics and information technology in economic development.[10]

While the TNCs provided opportunities, the source of Taiwan's explosive growth in electronics was local companies. These sub-contractors inundated the TNCs with offers of manufacturing services. Many Taiwanese technicians gained work experience and later left the TNCs, setting up their own businesses to supply market niches and services, sometimes to their former employers. Larger firms such as Tatung supplied the TNCs as sub-contractors and then learned to imitate and compete with them. As in the ROK, local entrepreneurs eagerly exploited opportunities for export growth, first in food processing, then textiles, and later electronics.[11]

Export orders from foreign buyers caused a rush of new domestic firms into radio production in the mid-1960s and, later, TVs, TV games and computers. Growth was stimulated as the larger electrical firms such as Tatung and Taco diversified into electronics in the mid-1960s (Chaponniere and Fouquin 1989 p. 30). Between 1966 and 1971 Taiwanese electrical and electronics exports together grew at an average rate of 58 per cent per annum (Wade 1990 p. 95).

During the start-up phase many Taiwanese companies learned the art of manufacture. They relied heavily on foreign firms for training and licensing agreements. Between 1952 and 1988 the government approved more than 3,000 such agreements; many were for technology transfer in electronics (Dahlman and Sananikone 1990 p. 78). Technology was also acquired by copying, reverse engineering and foreign training and education. Later on, many Taiwanese nationals returned from abroad after studying and working in foreign companies, a reverse of the brain drain. What began to be called the brain gain in the late 1980s was a significant source of new skills.

Foreign and local buyers were important sources of technology as many large and small American and Japanese companies bought large quantities of low-cost consumer electronics for resale in their home markets. As in Hong Kong and South Korea, the buyers frequently supplied product specifications, manufacturing process information and training to the local companies, not to mention export orders.

Industrial Take-off (*circa* 1970s)

During the 1970s consumer electronics took off, computers started up and some firms began designing semiconductors. Successive waves of product innovations in the West were exploited by latecomer firms, often in alliance with foreign TNCs and buyers. New product lines included colour TVs, digital watches, calculators, push-button telephones and TV video games.

Some local firms mastered the production technology for these goods but many continued to rely on OEM and foreign buyers for technical assistance. In the early 1970s, IBM began purchasing large quantities of sub-assemblies and components from Taiwanese companies.[12] Other US computer firms followed helping local companies learn about an increasingly sophisticated array of components and systems.

The government attempted to deepen industrial development by controlling and, in some cases, discouraging labour-intensive TNC investments, and by imposing more export and local content targets on foreign companies. Encouraging jointly-owned ventures for components, the government protected the local market and strictly controlled imports. Higher value-added production was encouraged by the lowering of taxes on selected technology imports, while tax write-offs for corporate R&D were introduced in the early 1970s.

By the mid-1970s electronics had become the second largest export industry after textiles. Many TNCs participated, including Philips which began making black and white TVs in 1970 and then colour TVs in Kaohsiung in 1976. Philips eventually became one of the largest TNCs in Taiwan, producing TV monitors, compact disc players and many other products. RCA, which began making memory circuits in 1969, started producing black and white TV sets and tuners in 1971. After 1976 it transferred metal oxide semiconductor (MOS) technology to local firms via the government's ITRI (interview ITRI, 1992).

Some SMEs entered at the technological state-of-the art in the 1970s. The best-known example is ACER, which was established in 1976 by Stan Shih and 11 engineers (mostly US trained) under the name Multitech International Corporation. By 1987 ACER's sales had reached US$331 million and by 1993 sales exceeded US$1 billion. By then it had set up an extensive network of outlets and manufacturing operations in China, the US and other countries (*Electronic Business* February 1993 p. 77).

Growth and Sophistication (1980s and 1990s)

The third phase, which saw rapid growth and technological deepening, took place during the 1980s and into the 1990s. Electronics output more than quadrupled in current prices, rising from around US$4.1 billion in 1980 to US$17.2 billion in 1990 (Table 5.1). Large numbers of small firms entered the professional electronics area, making computers, sub-assemblies, monitors, printed circuit boards, printers and keyboards. Latecomer firms consolidated their expertise in chip design while some companies (e.g. UMC) began to make semiconductors in medium volumes, relying on technological support from the government-funded ITRI.[13] To encourage technology deepening, the state invested in its own research laboratories and organized collaborative R&D ventures among local firms.

By the late 1980s wages had risen markedly and Taiwan's competitive advantage had progressed from cheap labour to low-cost, productive, high-quality engineering. As in the other dragons, production required increasingly complex precision engineering and electro-mechanical interfacing.

The PC industry took off in the 1980s as IBM, Wang, Hitachi and others purchased huge quantities of finished goods and sub-assemblies. By 1990 the export value of computers and related goods was more than double that of consumer electronics. Often through companies in Hong Kong, Taiwanese firms transferred much of their low-end production into neighbouring regions of China, stimulating the rapid growth of the Chinese economy.

The US technology connection deepened during the 1970s and 1980s. In 1978 Taiwan's national science advisor (also the chairman of TI), Dr Pat Haggerty, led a five person consulting group to Taiwan to advise on how to accelerate technology-intensive industries in Taiwan. The group recommended a thrust into semiconductors and computer technology. As a result, in May 1979 the Executive Yuan promulgated the Science and Technology Development Program which identified information technology systems as a key area. It also recommended the establishment of the Institute for Information Industry (III).

In July 1979 the MOEA contracted out the implementation plan for computer technology to ITRI and invited relevant state agencies and firms to jointly establish the III to promote the domestic use and development of computer systems (Executive Yuan 1989 p. iii–ive). A ten year development plan was prepared by the CEPED, covering the period 1980 to 1989. This established targets for computer use, R&D spending and skilled manpower supply. The Electronic Research Services Organization (ERSO) of ITRI was given the job of promoting R&D in computers and coordinating the transfer of technology from companies such as IBM, Microsoft and AT&T (Executive Yuan 1988 p. 11).

In the latter part of the 1980s, companies increased their exports of precision engineered goods such as hard disk drives, colour display terminals, video graphic adaptors, TV monitors and computer peripherals. Taiwanese firms increasingly competed at the early phase of the product life cycle, introducing new improved designs in anticipation of market needs. Compared with the *chaebol*, Taiwanese companies targeted high value-added niche markets, rather than scale-intensive production.

Semiconductor design and manufacture also took root. Philips began a joint venture with the Taiwanese Government in 1987, forming the Taiwanese Semiconductor Manufacturing Corporation (TSMC) to make specialist circuits for local design firms.[14] In 1991, TI and ACER formed a partnership to make memory circuits for the local computer industry. The fast-growing demand for microprocessors led Mitac (Taiwan's second largest computer maker) to discuss an alliance with Intel of the US in 1992 to make the 80586 central processing chip (*The China Post* 31 July 1992 p. 9).

By the early 1990s Taiwanese firms had more than proved their reputation as innovative designers of finished PCs, notebook computers and printed circuit boards. Tiny latecomer firms, mostly unknown in the West, grew to become fairly large companies. Datatech, for instance, which was founded 1981 by four friends from college, began assembling printed circuit boards in a converted apartment building (Johnstone 1989 pp. 50–51). By 1993 the company sold more than US$200 million worth of motherboards, PCs and clones of Sun Microsystems workstations and operated out of a modern, automated factory.

In semiconductors, after growing at rates of around 50 per cent per annum, Taiwan surpassed Britain to become the fifth largest producer worldwide in 1993. One of the industry leaders (UMC) boasted that Taiwan would soon become the world's third largest semiconductor producer (*Electronics* 14 February 1994 p. 9).

During the 1980s thousands of Taiwanese nationals went abroad to study and to work in foreign corporations. In 1988 more than 7,000 Taiwanese studied abroad (30 per cent or so in engineering), mostly in the US and Japan (Dahlman and Sananikone 1990 pp. 87–8). Through the 1980s locals trained in foreign TNCs became a direct source of technology. Entrepreneurs were attracted back by government incentives and the lure of high rewards. In 1991 nearly 1,000 engineers and scientists returned to Taiwan from the US

(Taiwanese Government data, cited in *Business Week* 30 November 1992 p. 76). The combination of returnees and government R&D investments led to several close-to-the-frontier innovative operations, especially in semiconductors.

5.5   STRATEGIC GROUPS IN TAIWAN

Using case examples, the following sections show how individual firms acquired technology and gained large shares of the international electronics market.[15] Compared with South Korea, strategic types of companies in Taiwan are more diverse and complex. The American connection also runs deeper than in the ROK, mostly because of Taiwanese corporate links with the US-dominated computer industry[16] and the proliferation of OEM and ODM partnerships.[17] However, in both countries, Japan became the controlling supplier of electronic components and capital goods due to its lead in electronic capital equipment and key components for low-end, consumer-type products.

Local industry is made up of at least five strategic types of firms: (a) foreign TNCs and joint ventures; (b) the major local manufacturing groups; (c) high-technology start-ups; (d) government-sponsored ventures;[18] and (e) the large numbers of traditional SMEs which clustered together to exploit market niches.[19] As a result of their success, some of the SMEs grew to become large international competitors during the 1980s, while others sank without trace. Entry, exit, flexibility and opportunism prevailed in the industry.

5.6   FOREIGN FIRMS AND JOINT VENTURES

Contribution of FDI

Although the scale and impact of FDI is rarely analysed, TNCs played a key part in Taiwan's industrial success.[20] Small in absolute terms, FDI contributed about 2.2 per cent of total domestic capital formation between 1965 and 1968. This rose to 4.3 per cent between 1969 to 1972 but fell off sharply to 1.4 per cent between 1977 to 1980, thereafter rising again to 2.5 per cent between 1984 and 1986.[21] Measured as a percentage of domestic capital formation in manufacturing, FDI oscillated between 2 per cent and 5.5 per cent between 1965 and 1983. These figures confirm that Taiwan, like South Korea, was relatively closed to FDI compared with Hong Kong and Singapore.

Although small, the contribution of FDI to industrial progress should not be underestimated. FDI, concentrated in leading export sectors such as electronics, gave rise to several new export industries, to the transfer of technology and eventually to the thriving OEM/ODM system. Foreign firms accounted for some 20 per cent of the country's total exports between 1974 and 1982. This fell to 16 per cent in 1985 as local firms assumed more importance. The TNCs accounted for around 16 per cent of those employed in manufacturing in 1975, a total of 245,000 people. This rose to 357,000 (17 per cent) in 1979 and then fell back to 234,000 (9 per cent) in 1985.

The electronics and electrical appliances sector benefited especially from FDI, accounting for around one third of Taiwan's total FDI up until 1974 (Wade 1990 p. 149) and for some 36.3 per cent of total investment in 1987. The next largest sector was chemicals which accounted for around 17.5 per cent. FDI helped stimulate Taiwan's electronics producers and raise the standard of exports. Since the mid-1980s the share of FDI in industrial output fell as local industry grew in stature and capability. Nevertheless, TNCs and joint ventures still transferred technology to local firms and linked them into large foreign markets.

As shown below, entrepreneurs exploited FDI to their advantage, eventually making the TNCs dependent on their manufacturing skills and their highly productive engineering talent. Local companies reversed the pattern of TNC exploitation, often complained of in Latin America and other developing areas, by supplying low-cost OEM/ODM services, components, subsystems and complete products. Indeed, as Section 5.11 shows, hundreds of local companies learned to innovate by clustering around the TNCs.

TNC Strategies

The first strategic group (TNCs and joint ventures) initiated much of the electronic industry and provided opportunities for latecomers to become subcontractors, OEM suppliers, licensees and, eventually, competitors. Among the major Western TNCs were Philips, RCA, IBM and DEC.

Philips

Taking one example, Philips of Holland played a leading part in developing the electronics industry in Taiwan, upgrading its production and technological facilities, working closely with local firms, and transferring electronics technology (Table 5.4). In return, the company benefited from Taiwan's rapid industrial growth and the highly skilled, low-cost labour force.

Philips's sales grew from around US$117 million in 1984 to US$276 million in 1987, while employee numbers increased from 3,000 to over 5,300 in the same period. By 1991 sales were in the region of US$970 million. In recognition of the importance of Philips Taiwan Ltd to the overall Philips Group, C.W. Cheng, a Taiwanese national educated in the US, was appointed to the position of executive vice president to take charge of Philips's Asian strategy (*China Post* 28 July 1992 p. 13).[22]

Table 5.4   Taking root: Philips's production and technology transfers (a selection)

1961	began labour-intensive packaging operations
1966	opened major facility in the Kaohsiung free trade zone (for resistors, capacitors and simple integrated circuits)
1970	began manufacture of cathode ray tubes
1975	purchased a black and white TV set assembly factory from a US firm
1978	began colour picture tube production
1986	formed a joint venture with Avnet in Hsinchu for compact disc readers
1987	started up a semiconductor design centre
1980s	transferred semiconductor technology to ERSO/ITRI
1987	formed a joint venture with Taiwanese Government (TSMC) to transfer static random access memory (SRAM) semiconductor technology.

Sources: Industry interviews, company reports, ITRI and secondary sources (see text).

As a leading investor in the free trade zones, by the late 1980s Philips had graduated from low-end products to top-of-the-range electronic goods including semiconductors, flat-screen TVs and liquid crystal monitors. A network of local companies grew up to supply Philips with components, subsystems and services. Some imitated Philips, competing in selected market niches. The company transferred process and product technologies and trained up local engineers, technicians and directors.

Philips's integration into Taiwan's industrial structure culminated in 1987 with an agreement to transfer chip technology to the Taiwan Semiconductor Manufacturing Company (TSMC), a joint venture between Philips and the Taiwanese Government. TSMC was to produce specialist chips for both local firms and Philips's own internal use. Under the venture, ERSO/ITRI arranged with Philips to transfer static random access memory (SRAM) technology used in consumer electronics. After training from Philips, ERSO spun off around 200 personnel, mostly engineers, to join the new company.

TSMC set up two leading-edge fabrication lines, both located at Hsinchu and later became one of the first companies in the world to offer foundry-only (i.e. fabrication with no design) services, mostly for Silicon Valley and Taiwanese chip design companies. By 1991 TSMC employed around 770 staff. In 1992 it expanded further by allocating around US$1 billion for a new chip plant, making it one of the largest semiconductor companies in Taiwan.

RCA

RCA, an early US entrant, also readily transferred technology and took root in Taiwan. Beginning with the assembly of simple memory chips in 1969, in the early 1970s it introduced black and white TVs, tuners, and monitors for export. As with Philips, growth in export demand led RCA to upgrade its engineering facilities, to transfer technology and to work closely with many local firms. By 1988 RCA employed some 2,600 local workers, of which 400 were engineers, technicians and administrative staff. Around 75 worked in the R&D department.

In 1976 RCA began one of the first ventures to transfer chip technology to local firms through ERSO/ITRI. Its American parent supplied ERSO with training in chip processing, design engineering, cost accounting, sales and marketing, new materials and operations management (interview, 1992). A one-year training programme in RCA in the US was designed for 30 of ERSO's engineers. One of these engineers later founded Winbond Corporation which, by 1992, had become the second largest chip producer in Taiwan, with 1,100 employees and sales of more than US$1.3 billion (see Section 5.10 below). In the US, the Taiwanese engineers learned to use RCA's special CAD systems for CMOS chip technology. The entire system was then purchased and shipped back to ERSO in Taiwan. US engineers then helped local staff to set up a pilot facility to manufacture ERSO's first CMOS chips, initially for the digital watch market. Several other ERSO spin-off firms benefited directly from RCA's training and technology transfer.

Sanyo

Sanyo Electric, an example of a major Japanese investor in Taiwan, began in 1963 by forming a partnership with a local trader. In contrast with most US firms, the Japanese initially supplied the local market and only later began exporting. By the late 1980s exports accounted for around one third of Sanyo's production.

Sanyo was one of the first companies in Taiwan to make modern white and brown goods. After introducing TVs in 1969, it began producing VCRs in the 1970s. During the 1980s it competed with Sony, Matsushita and Sharp, as well as Tatung, to supply the domestic market for TVs, VCRs and white goods. By 1988 Sanyo had won 10 per cent of the local TV and VCR markets and around 20 per cent of the domestic refrigerator market. As a result of rising wage costs, Sanyo ceased expanding after the late 1970s. The workforce declined from

around 3,800 in 1984 to around 3,400 in 1987 as it relocated production to
China and Thailand (Chaponniere and Fouquin 1989 pp. 47–8). By the late
1980s around 90 per cent of the parts and components for Sanyo's TVs and
VCRs were purchased locally, many from local Taiwanese companies.

Like US firms, Japanese corporations acted as pacing horses for latecomer
firms. They stimulated competition in consumer electronics and white goods,
and helped create a thriving network of local suppliers of parts and services.

5.7   THE MAJOR MANUFACTURING GROUPS

The second strategic group, Taiwan's major manufacturing conglomerates,
also forced the technological pace of local industry. As Chaponniere and
Fouquin (1989 pp. 37–9, p. 55) show, the groups share several common
features. They are all fairly old and diversified and initially developed to
serve the local market.

Tatung
Tatung, the largest electronics maker in Taiwan, was the fourth biggest industrial group in Taiwan in the late 1980s, ahead of Sampo, Teco and AOC.
Tatung was founded in 1919. It diversified from construction into building
materials in 1926, mechanical equipment in 1939, heavy electrical equipment
in the 1940s and electric fans in 1949. The company also moved into steel,
chemicals and telecommunications. In 1985 Tatung's electronics sales reached
around US$640 million. It had subsidiaries in the US, Japan, Singapore,
Hong Kong, Germany, Thailand, Indonesia and South Korea.

During the 1970s, electronics became Tatung's largest operation. The
Group's steady progress up the technological ladder mirrors the development
of the overall electronics industry in Taiwan. As Table 5.5 shows, the company advanced from black and white TVs in 1964 to high-resolution colour
TV tubes, disk drives and semiconductors in the late 1980s.

Like the South Korean *chaebol* Tatung assimilated manufacturing know-
how initially under technical cooperation deals, then by licensing and OEM.
It began by acquiring relatively mature process technologies for household
appliances and consumer electronics from both US and Japanese companies
through technical assistance deals and by investing capital in joint ventures
with foreign companies.

The company learned many of its technological skills under OEM deals.
By the late 1980s, around half of its colour TVs, PCs and hard disk drives
were exported under OEM. Most production embodied little original R&D,
but the company had closed much of the process and product technology gap
with leaders in advanced countries. Tatung learned the ability to absorb and

Table 5.5 Tatung's progress in electronics

Product	Introduction date
Black and white TVs	1964
Colour TVs	1969
Black and white TV picture tubes	1980
VCRs	1982
High resolution colour TV picture tubes	1982
Colour TV tubes	mid-1980s
PCs	mid-1980s
Hard disk drives	mid-1980s
TV chips/ASICs	late 1980s
Sun workstation clones	1989
Fourteen-inch colour monitors	1991

Various sources.

adapt advanced foreign technology and to modify, re-engineer and re-design consumer goods for different types of customers, especially those in East Asia. Its in-house engineering capabilities were used to scale down production processes, adjust capital to labour ratios, and to bring about continuous improvements to production technology.[23] By 1990 Tatung employed around 500 R&D staff in electronics although most of their work was in advanced engineering rather than blue sky research.

By the mid-1980s Tatung had the confidence to transfer production technology to lower cost East Asian countries, including Thailand and the Philippines. Beginning with mature process technologies for low-end goods, Tatung's subsidiaries benefited from the parent's long experience of harsh developing country conditions in Taiwan.

Tatung first invested in advanced countries in the early 1970s, following initial export success and the need to set up sales offices to service local markets. Later, overseas production helped Tatung to compete in foreign markets, establish and reinforce its brand image and circumvent European quota restrictions. In 1978 it acquired a TV manufacturer based in Los Angeles, and another in 1980 in Atlanta. In the same year it acquired Decca (in the UK). By 1981 Tatung had eight manufacturing operations abroad making TVs, TV picture tubes, electric fans, washing machines, refrigerators and other household electric goods. Since then it has set up plants in Ireland and Singapore.

5.8   FROM OEM TO ODM TO OBM

Although each of Taiwan's industrial family groups has its own particular history, each acquired technology and learned to innovate incrementally, often by imitation. TECO, another major group, learned under OEM arrangements with IBM and other companies. TECO, like Tatung, advanced from simple consumer goods to computers, colour display terminals, printers, video graphic adaptors and TV monitors. By the late 1980s TECO electronics sales exceeded US$300 millions, employment stood at over 3,000 and R&D staff numbered in the region of 300.

Like many of Taiwan's SMEs, the major groups began with OEM arrangements because they lacked their own marketing capabilities and brand names. Table 5.6 captures the progression of Taiwan's latecomers, from OEM, to ODM and, in some cases, to own-branded goods. By the early 1990s, like the *chaebol*, several of the groups had established OBM in at least some areas, although most still depended on OEM/ODM for the majority of their exports.

Table 5.6   Transition of latecomer firms: from OEM to ODM to OBM[a]

	Technological transition	Market transition
OEM	Learns assembly process for standard, simple goods	Foreign TNC/buyer packages, brands and distributes
ODM	Local firm designs[b] Learns product innovation skills	TNC buys, brands and distributes TNC gains PPVA[c]
OBM	Local firm designs & conducts R&D for complex products	Local firm organizes distribution, own-brand name and captures PPVA[d]

Notes:
[a]   Applies to electronics and other fast growing export markets, such as footwear and bicycles and to firms in the other dragons as well as Taiwan.
[b]   Post-production value-added.
[c]   Or contributes to the design, alone or in partnership with the foreign company.
[d]   In bicycles Taiwanese firms made the full transition to OBM, unlike electronics.

Source: See text.

5.9 HIGH TECHNOLOGY START-UPS

Taiwan's Unknown Latecomers

In contrast with the industrial groups, many of Taiwan's high technology start-ups entered in the late 1970s and early 1980s with product innovation/ ODM capabilities, often gained by individuals with overseas experience in US firms or universities. Many continued to rely on OEM to some extent and most were virtually unknown outside of East Asia, despite some brand name sales. Most began as niche market players and remained focused on a narrow range of products.

Of these companies ACER (examined below) was the best known outside of Taiwan in the early 1990s. However, there were many other highly respected firms in the country. First International Computer Inc., the world's largest producer of circuit boards for PCs in the early 1990s, had sales of US$230 million in 1993 and plans of becoming a US$4 billion corporation by the year 2000 (*Business Week* 28 June 1993 pp. 36–7). First International formed joint product development ventures with leading US firms such as Intel, TI and Motorola to add to its technical abilities.

Datatech Enterprises Co., mentioned in Section 5.4, was one of the largest international motherboard producers, selling more than US$200 million in 1993. Another leading latecomer, Elitegroup Computer Systems Corporation, claimed a 10 per cent world market share in motherboards in 1993 (*Electronics* 26 April 1993 p. 12).

It is impossible to mention all the significant latecomers. In computers, around 20 local firms produced 54 per cent of Taiwan's output in the late 1980s, leaving the other 46 per cent to hundreds of SMEs, many of them newly formed, focusing on specialist niches. In 1989 alone at least 30 new firms began laptop production, adding to the rivalry and dynamism of the industry. One of the larger latecomers was Mitac, ACER's arch competitor in computers. Mitac boasted an annual PC turnover of more than 200,000 units in the latter part of the 1980s.

Another latecomer, Cal-Comp, virtually unknown in the West, was the largest producer of calculators worldwide in 1992 and Taiwan's largest fax machine maker (*China Post* 28 July 1992 p. 9). Under OEM/ODM, it made roughly 80 per cent of Japanese Casio calculators. Many of Japan's leading fax machine makers established business links with Cal-Comp to gain from its high-quality, low-cost, mass-produced electronics. Another barely known was Twinhead, which sold around US$160 million worth of notebook computers in 1992, some under its own brand name.

The Case of ACER

ACER is interesting because it illustrates many of the contemporary strengths and weaknesses of Taiwan's latecomer corporations. As mentioned earlier ACER started with 11 engineers in 1976. By 1993 its sales had reached US$1.4 billion. Leading the local computer industry in the 1980s, ACER constantly strived for own-brand recognition abroad. By 1988 60 per cent of its sales were own-brand (Johnstone 1989 p. 51). However, in 1992 the firm scaled down its OBM efforts and returned to OEM/ODM after sustaining heavy losses.

During the late 1980s, ACER was among the world's largest producers of PCs, colour monitors, keyboards, fax machines and printers. It developed its own-brand workstations, operating systems and new chip designs. In 1992 ACER further diversified into memory integrated circuits through a joint venture with TI, contributing around US$70 million to a US$400 million facility.[24]

After its incorporation ACER grew rapidly as its technology flourished. By 1988 company turnover was around US$600 million, including 400,000 PCs (around 6 per cent of the world market). In 1988 the ACER Group employed roughly 4,000 staff, of which 500 or so worked in R&D. This increased to around 800 in 1992.

As a major OEM supplier to ITT, AT&T and other market leaders, ACER consistently showed original innovative capabilities. It designed the first Chinese operating system (called Dragon) which later became a standard in Asia. With IBM, Apple and several Japanese companies it helped define Asian computer standards.

Table 5.6 presents a small number of ACER's achievements. Like many other Asian latecomers, ACER developed systems and software innovations upon a solid foundation of process technology. On the one hand, the company developed new software capabilities from behind the R&D frontier set by Intel, IBM, AT&T and others. On the other hand, these efforts were funded by revenues gained from low-cost manufactured goods and sales under other company brand names. In 1993 up to 40 per cent of ACER's output was still sold under OEM/ODM.

As Table 5.7 shows, ACER produced significant innovations, many of which were highly valued by users. Ultimately the company turned the tables, licensing back technology to leaders such as Intel. Many of ACER's own-brand goods were intelligently modified and improved PC designs.

During the 1980s and early 1990s, most of ACER's production was for export, evenly split between the US and Europe. Exports sold to 70 different countries through a retail network of around 100 distributors. Offshore manufacturing plants were set up in the US, Holland, Malaysia and China. After

Table 5.7   ACER – behind the frontier innovations: a selection

1984	developed its own version of the 4 bit microcomputer (later followed by 8 bit, 16 bit and 32 bit PCs)
1986	launched the world's second 32 bit PC, after Compaq but ahead of IBM
1988	began developing supercomputer technology using the Unix operating system
1989	produced its own semiconductor ASIC to compete with IBM's PS/2 technology
1991	formed a joint company with TI (and the Taiwanese Government) to make memory chips (DRAMs) in Taiwan
1992	formed alliances with Daimler Benz and Smith Corona to develop specialist microelectronics technology
1993	produced a novel PC using a reduced instruction-set (RISC) chip running Microsoft's Windows NT operating system
1993	licensed its own US-patented ChipUp technology[a] to Intel (in return for royalties)
1993	received royalties from National Semiconductor, TI, Unisys, NEC and other companies for licensing out its PC chipset designs.

Note: [a] Allowed a single-chip upgrade to a dual-Pentium microprocessing system.

Source: See note 24.

purchasing Altos in 1980 to distribute computers directly into the US, in 1987 ACER took over Counterpoint, an American supplier of minicomputers. By 1993 its US plant (ACER America) employed around 500 people and produced roughly 16,000 PCs a month.

In an attempt to challenge brand leaders and move beyond OEM, ACER began to distribute its own brands directly to customers in the US and Europe, while the company's founder and managing director, Stan Shih, began Taiwan's Brand International Promotion Association in an effort to build up Taiwan's quality brand images abroad. Despite progress in design and branding, ACER still relied on OEM for around 50 per cent of its monitor sales and 20 per cent of PCs in 1992.

After losing in the region of US$90 million between 1990 and 1993, ACER retreated from OBM to return to traditional OEM and ODM sales. In 1993 the Group reported a profit of US$30 million, partly as a result of the strategy. Focusing on OEM/ODM allowed ACER to increase output and gain economies of scale and customer-led quality improvements. It also helped reduce further heavy marketing and distribution investments needed to increase OBM sales. Under the OEM strategy, ACER negotiated one contract worth around US$100 million to supply Apple Computer Inc. with a popular notebook computer (the PowerBook 145).

To sum up, ACER, like other high-technology start-ups, benefited from Taiwan's improving technological infrastructure and established export market channels. In contrast with the first latecomers, ACER was able to enter at a level closer to the technology frontier set by leading TNCs, avoiding the 1970s phase of consumer electronics. ACER quickly began to innovate simultaneously with software, products and processes for manufacture. Several key engineers and managers benefited from foreign education and from working in US corporations. ACER's rapid moves from OEM to OBM were supported by a strong in-house R&D effort.

ACER also illustrates the vulnerability and weaknesses facing many of East Asia's latecomer firms. Against its desire, ACER was forced to retreat from own-brand sales after sustaining heavy losses. Early OEM deals allowed ACER to grow by working as a sub-contractor to Intel and others, but this path limited ACER's strategic room for manoeuvre. Under OEM, ACER remained dependent on leaders for core technologies, market outlets and better-known brand names. Its retreat back to OEM after the foray into OBM

Table 5.8   High-technology start-ups in Hsinchu Science-Based Industrial Park

Firm	Start date	Sector	Sources of senior staff, technology and training[a]
Microelectronics Technology Inc.	1983	Telecom	HP, Harris, TRW
United Fibre Optic Communications Inc.	1986	Telecom	Sumitomo (J), Philips (H) AT&T, STC (UK)
TECOM	1980	Telecom	Bell Labs, IBM
Macronix	1989	Semiconductors	Intel, VLSI-Tech.
Winbond Electronics Corp	1987	Semiconductors	RCA, HP
Taiwan Semiconductor Manufacturing Corp.	1987	Semiconductor foundry	Harris, Burrows RCA, Philips (H), IBM

Note: [a] All US firms unless indicated H=Holland; J=Japan: UK=United Kingdom

Source: Company interviews carried out in 1992

shows how difficult it is for companies to overcome their latecomer orientation and disadvantages. Many of the larger latecomers interviewed believed that unless and until they made a full transition to OBM and in-house technology, they would remain subordinated to the international leaders and unable to compete on equal terms.

5.10 GOVERNMENT-SPONSORED START-UPS

The fourth strategic category, government-sponsored start-ups, overlaps with the third, but differs in that there was more direct assistance from ERSO/ITRI to overcome technological barriers to entry. Table 5.8 presents a sample of such firms, interviewed at the Hsinchu Science-Based Industrial Park in 1992. Each had received direct or indirect assistance from ERSO/ITRI and/or each had benefited from the Hsinchu industrial park facilities. Directors from each company had been trained in the US, many in Silicon Valley. Some directors maintained training and licensing agreements with their former US employers. This group shared many of the characteristics of the other high-technology start-ups, with firms combining OEM, ODM and OBM.

Microelectronics Technology Inc. (MTI)

MTI, a telecommunications equipment maker, produced integrated circuits for microwave systems as well as components for direct broadcast satellite, digital radio and maritime systems. MTI started in 1983 with eight founders, all from Silicon Valley. By 1992 company turnover had risen to around US$100 million, employment was in the region of 800, of which 700 worked in Taiwan and a further 100 in the US. Each of the founders had worked for US corporations, including HP, TRW and Harris. They returned to Taiwan partly in response to the government's programme to attract back overseas Taiwanese and partly because of the profit opportunities. The government's package included tax incentives, loans and the use of science park facilities at Hsinchu.

The skills and connections of the founders meant that acquiring technology presented little problem to the company. MTI began with a small line of integrated circuits for microwave niches, mainly receiving dishes for satellites and systems for ships and aircraft. Among its initial customers were the government and local satellite broadcasters.

MTI's first large export orders were produced under OEM for American manufacturers who then supplied satellite TV broadcasters in the US. The early products were labour-intensive and the attraction to US buyers was low cost. Working with component suppliers from the US and Japan, and OEM clients, the directors learned new technological and marketing skills.

With expansion, MTI's output became progressively automated, while manufacturing techniques were learned under OEM. By the mid-1980s the firm began to introduce new product innovations, some of which were offered to customers under MTI's own brand. In 1986 a new special low-noise amplifier for the direct broadcast satellite field was developed, along with the world's smallest and lightest INMARSAT standard-A terminal for maritime communications. Several other product innovations helped the company to grow and improve its marketing image in the US.

In 1992 around 90 per cent of products were exported, of which more than half were still supplied under OEM/ODM. Regarding OBM strategy, the company's small size preempted direct competition with major suppliers and constrained large-scale R&D spending for radical innovations. However, small size brought advantages. Like other local firms, MTI was able to respond quickly to changing market niches.

As of 1992, MTI's manufacturing facilities were comparable with those of leading firms worldwide. Facilities included CAD/CAM workstations for designing microwave circuits, clean room facilities, precision equipment for mask aligning, automatic dicing systems and so on. In some cases the equipment was the latest available. Of the 700 employees, around 130 were engineers with degrees, including some Ph.D.s. Of the balance around 50 per cent were educated at local universities.

In order to supply overseas markets and carry out joint R&D work with major clients, MTI set up two subsidiaries, one in the US and one in Canada. The primary customer base in 1992 still reflected the origins of the company. Former employers of the founders, HP, Digital Microwave Corporation and Hughes Network Systems were major clients. Other customers included STC Marine and BT in the UK. After starting up, MTI moved rapidly to combine OEM with ODM and OBM in selected areas. Government support helped initiate the firm, but its subsequent strategy was an internal matter.

Winbond Electronics Corporation

A second example, Winbond Electronics Corp., illustrates a different type of company, oriented towards the local market (as well as exports), rather than exclusively towards the export market. Internal markets demanded a different innovation strategy, with the government (through ERSO) arranging technology transfers. The case illustrates strengths and weaknesses of the new start-ups and shows the advancing status of the local electronics market. Unlike OEM suppliers, Winbond began competing at the leading edge of chip technology. The company strategy was to exploit the explosive local demand for components and the technological demands of domestic computer makers.

Winbond started in 1987 as a PC chipset manufacturer. By diversifying into memory devices and then into ASICs, by 1992 the company had become the second largest local chip maker, employing around 1 100 staff, including 65 production engineers and 55 design engineers. In 1992 sales were US$130 million, the majority of which (around 60 per cent) were destined for local PC makers. Around 10 per cent of turnover was invested in R&D in 1992.

Tracing the origin of the company, the founder studied solid state physics at Princeton from 1970 to 1974. In 1975 he joined ERSO/ITRI, where he worked for 11 years. One of his tasks was to arrange chip technology transfer from RCA in the US, where he spent one year on a training programme. As discussed in Section 5.6 above, RCA trained ERSO/ITRI's production, design, marketing and operations engineers. In 1987 ERSO/ITRI decided to terminate commercial chip operations and to spin off the activity to the private sector. Initially eight experienced engineers left to form Winbond. The eight were followed by 30 or 40 more ERSO/ITRI employees.

Having raised starting capital Winbond agreed to license the CMOS chip technology from ERSO/ITRI, also acquiring a ready made customer base. Within 11 months, the new company had built a semiconductor factory and had begun manufacturing. The principal investment financier was Walsin Cable which took a 50 per cent share in the company. In 1987 the Taiwanese capital market was booming and other investors were keen to supply capital.[25]

After incorporation, the numbers of customers grew as Winbond developed its own design and development capabilities, often in cooperation with US firms. In 1992 most development work was carried out in Taiwan, though some design work happened in the US with larger customers. Joint work included a project with NCR's Microelectronics Division in Colorado for gate array development.

In 1992 Winbond's facilities were at the standard of market leaders in other countries. In 1988 it installed a semiconductor fabrication line for medium volume, mainstream (1.2 to 1 micron, 5 inch wafer) technology which included a class 10 clean room and state-of-the-art steppers, dry etchers and ion implanters. A second fabrication line was under construction in 1992 for six-inch wafers (0.6 to 0.8 micron, CMOS process capabilities), costing around US$257 million. The strategy was to produce mainstream (16 megabit DRAMs and 4 megabit SRAMs), but not quite leading-edge chips. The normal supporting facilities (e.g. computer-aided design systems and SPARC workstations) were installed by an experienced in-house engineering/R&D team recruited from the US and Taiwan.

In product technology, Winbond matched the design ability of Silicon Valley companies with its own complex cell based ASICs and full custom integrated circuits. Although most new design engineers were recruited from local universities, a few senior engineers were from US firms. Licensing, still an important part of Winbond's operation in 1992, included HP's PA RISC chip (an advanced microprocessor) to which it added new design features suited to the low-priced, high-volume terminals and printers made in Taiwan. In return HP gained Winbond's knowledge of the needs of Chinese-speaking users, not only in Taiwan but also in Hong Kong, Mainland China, Malaysia and Singapore.

Winbond illustrates the typical strengths of Taiwanese start-ups: speed, agility, market responsiveness and entrepreneurship. Seizing new opportunities in the local and export markets the company used the ERSO/ITRI connection to its advantage, absorbing foreign technology quickly and learning to innovate with new products in the mainstream of technology.

However, the company suffered from many of the constraints of small-scale operations: shortages of investment finance; poor brand name recognition; uncertain distribution arrangements. Winbond, like many other latecomers, remained dependent on US firms for leading-edge designs and important capital goods. As of the early 1990s, Winbond still relied to some extent on OEM/ODM, despite its innovative skills.

United Fibre Optic Communications Inc.

A third example of a government-sponsored start-up is United Fibre Optic Communications Inc. (UFOC). After beginning in 1985, UFOC became the first local firm to enter the market for fibre optic transmission cables. By 1992, turnover was around US$40 million and employment roughly 100. The idea to create UFOC originated within the MOEA Industrial Development Bureau. Believing that Taiwan needed an indigenous fibre optic producer the MOEA called together the four largest copper cable companies and the local telecommunications operator. Together they began UFOC as a joint venture company.

The founder of the company, an engineer, worked for the state telecommunications department for 32 years and then become the president of one of the investing cable companies. The two cable suppliers each purchased a 30 per cent share of the company. A further 20 per cent was owned by banking investors and another 20 per cent was held by employees and other individuals in 1992.

To acquire technology, UFOC asked four companies to tender (AT&T, Sumitomo, Philips and STC). In 1985 a licensing agreement was formed with AT&T in Atlanta under which manufacturing began in 1986. Some engineers were sent to AT&T for training while others were hired from the telecommunications operating company and the cable manufacturers. AT&T's technology was, in the first phase, mature and well proven. UFOC was able to learn by close cooperation under the licensing arrangement, once production had started. Thereafter, local engineers gradually took over the responsibility for running the operation, learning to resolve technical problems on site.

However, having absorbed the basic process technology, UFOC discovered that it could not upgrade without the assistance of AT&T. Strategically, the firm would have preferred in-house control but, as with many latecomers, it was forced to follow the technological lead set by the licenser. The licensing contract also prevented UFOC from selling directly into the international market.

In 1992 two strategic options faced UFOC. The first was to continue purchasing know-how under licence (although few companies would part with the latest technology). The second was to invest heavily in its own technology. Both options had advantages and disadvantages. The first would restrain the company, particularly on the international market. The second would be risky and costly and could jeopardize existing licensing business.

Although the firm had not yet reached a strategic decision on how to proceed, UFOC illustrates several interesting features of Taiwan's latecomer progress, not least the willingness of public sector employees to turn their hands to risky business ventures (as with Winbond). The case confirms how learning continues to rely on foreign TNCs and highlights the continuing strategic dilemma of latecomers, who find themselves overly dependent on international competitors for technology. Like the other cases, UFOC fits into the advanced stages of the marketing-technology model in Chapter 3 and illustrates the difficulties facing firms which attempt to overcome the sub-contracting path.

5.11 A MODEL OF INDUSTRIAL CLUSTERING IN TAIWAN

This section goes beyond single-firm strategies to analyse the dynamics of multi-firm clustering in Taiwan. It shows how whole groups of latecomer firms exploited new opportunities provided by TNCs and foreign buyers. Unlike the conventional user–producer clusters described by Vernon (1960), Lundvall (1988) and Porter (1990), latecomer clustering occurred behind the innovation frontier, largely in the absence of local leading-edge users and related industries.[26] To illustrate the nature, origin and growth of Taiwan's clusters, examples from electronics, footwear, sewing machines and bicycles are compared. Although broadly similar patterns are evident, differences in the strategies followed by both latecomer suppliers and foreign partners significantly affect the observed results.

Computer Keyboards

Taiwan's computer keyboards exports grew from humble beginnings in 1983 to around US$80 million in 1987. By that time there were 44 keyboard suppliers listed in the TEAMA buyers guide. Typically, latecomer firms began with fewer than 20 employees with well below US$1 million annual turnover (Levy 1988 p. 48). Some firms were started up by young graduate engineers with little industrial experience while others diversified in from other sectors.

Not unexpectedly, the latecomers suffered from capital shortages and their tiny size. However, as Levy (1988) shows, their adaptiveness and speed enabled them to overcome barriers to entry. Typically, firms would begin by supplying a TNC, foreign buyer or local trader with parts or components. The initial strategy was to focus on services and market niches with low initial start-up cost. Information on new opportunities was supplied by hundreds of traders operating from Taipei.

Many local producers grew larger during the 1980s, often giving rise to new start-ups *via* sub-contracting, imitation and spin-offs. Tracking a sample of three keyboard makers from 1982 to 1987, Levy (1988) shows how one grew from US$1 million to US$9 million in sales; the second from US$4 million to US$21 million; the third from US$2 million to US$36 million.

Supplying the TNCs and traders in volume, the three keyboard makers purchased from existing firms and encouraged the entry of new ones. The three bought keycaps, keyswitches, printed circuit boards, plastic cases and other components. When demand rose unexpectedly, they would contract out assembly, or parts of the assembly process, to expand capacity and ensure orders were met on time. One activity frequently contracted out by many companies was the mounting of electronic components onto printed circuit boards. This gave rise to Taiwan's printed circuit board industry which, by the early 1990s, was the largest in the world.

The Linkage-Clustering Model

In keyboards, *backward linkages* were forged with many new suppliers. Many local companies learned new skills, resulting in a low-cost, highly responsive industrial infrastructure which, in turn, attracted more foreign buyers and TNCs. Following Hirschman (1958) and Schive (1990), these further foreign investments can be called the *forward linkage* effect.

Carrying on the process, the forward linkage effect itself encouraged many more entrants forming a further, much larger, backward linkage effect. Successive waves of forward and backward linkages ultimately created the industrial clusters in keyboards, PCs, consumer electronics, computer mice, fax machines, calculators and other product areas. Requiring little R&D, and

> > > Initial Foreign Investment/Purchase < < <
Foreign firm enters in search of cheap labour

> > > > Backward Linkage Effect < < < <
Local sub-contractors enter to supply pieces, parts and assembly services

> > > > > > Forward Linkage Effect < < < < < <
More foreign investors/buyers enter to exploit low cost suppliers

> > > > > > Second, Larger Backward Linkage Effect < < < < < <
More local firms enter, many grow larger, infrastructure improves

> > > > > > > Backward–Forward Linkage Repeated < < < < < < <
Process is repeated in waves of foreign purchasing and local entry

> > > > > > > > Industrial Cluster Created < < < < < < < <

Note: *a* A similar process occurred in Singapore among TNCs in the disk drive industry (see Chapter 6). In Hong Kong in the early 1990s the backward linkage industries employed around half the number of the electronics industry. In South Korea, the process of growth was internalized within the *chaebol*, rather than in sub-contracting networks.

Figure 5.1   Industrial clustering from behind the frontier in Taiwan[a]

occurring largely under sub-contracting and OEM arrangements, this process of industrial clustering occurred behind the technology frontier set by the market leaders. Figure 5.1 presents a simple model of the process.

Personal Computers

Turning to PCs, exports grew from around US$12 million in 1983 to roughly US$393 million in 1987, by which time there were around 120 local manufacturers (Levy 1988 p. 48). Two PC makers began with sales of below US$1 million and less than 20 employees. In both cases by 1988 their turnover exceeded US$100 million. The two followed emerging market trends and a similar pattern of backward–forward linkages occurred as in keyboards.

Many small PC makers won their first orders from traders in close touch with new export markets. Within a few years, the larger companies tended to bypass the traders and sell directly to foreign buyers and overseas manufactures. The low-cost, adaptable OEM suppliers proved irresistible to foreign buyers and TNCs who expanded their purchasing operations in Taipei during the take-off period of the mid-1980s. In the period 1985 to 1986 the number of computer and software start-ups increased by nearly 50 per cent to around 151 in Taipei alone, while employment rose to around 38,000 (III 1988 p. 37).[27] By 1989 there were around 650 computer exporters and thousands of traders (III 1991 p. 37).

The dynamism of Taiwan's clusters was partly due to currency changes. With the appreciation of the Japanese Yen against the New Taiwanese Dollar production of many low-end goods became less profitable in Japan and OEM buyers increasingly sourced from Taiwan (III 1988 p. 39). In computer products OEM accounted for some 43 per cent of Taiwanese production in 1989, compared with 35 per cent by foreign-invested companies and 22 per cent for local brands.

Large foreign TNCs eased Taiwan's entry into the computer industry. IBM alone purchased around US$1.3 billion worth of computer products during the period 1966 to 1987 (Chaponniere and Fouquin 1989 p. 44). In the early 1970s, IBM bought huge quantities of sub-assemblies and simple components, exploiting Taiwan's low-cost labour. By the early 1990s, IBM and other leaders depended upon Taiwan's high resolution monitors, keyboards, printed circuit boards, graphics cards and printers. The six largest foreign buyers were (in order): IBM, Philips, NEC, Epson, HP and NCR (III 1991 pp. 39–43).

The initial TNC motivation of low-cost labour was transformed as latecomers augmented their engineering process and product design skills. By the early 1990s, most of the world's leading computer brands embodied products made in Taiwan. Table 5.9 shows OEM buyers and their local suppliers for desk-top PCs, monitors, notebook computers and motherboards

Table 5.9   US OEM purchasers and Taiwanese suppliers of computer products 1993

US Brand	Taiwanese supplier	Product
Apple	Acer	Power-book 145
Packard Bell	Tatung	Desk-top PCs
Dell	Inventec	Notebooks
IBM	Datatech	Motherboards
AST	Compal	Monitors
NCR	First International	Motherboards
Zenith	Inventec	Notebooks
Compudyne	Twinhead	Notebooks

Source: Company reports, amended from *Business Week* (28 June 1993 p. 37)

in 1993. In each product case, the forward–backward linkage effect had given rise to industrial clustering in Taiwan.

Sewing Machines

Taiwan's other fast-growing export industries engaged in behind-the-frontier clustering. In footwear, bicycles and sewing machines, groups of latecomer firms exploited foreign technology channels, competed fiercely and began catching up technologically. As in electronics, firms began with low-cost manufacturing and graduated to product innovation through time. By linking export market outlets to sources of learning, firms grew under sub-contracting relationships similar to OEM and ODM.

Taiwanese sewing machine makers transformed a tiny backward sector into the largest supplier in the world. A small initial investment by a TNC (the American-owned Singer Company) gave rise to the rapid diffusion of technology and to the start-up of a host of SMEs. Schive (1990, ch. 5) records in detail how the clustering process occurred.

Singer was set up in 1963 with a capital outlay of around US$0.8 million. Around 250 local firms made parts and machines for the local market but they lacked scale and quality was poor. The government negotiated Singer's entry to try to reduce imports of finished machines, create jobs and to save foreign exchange. The local content agreement with Singer stipulated that over 80 per cent of inputs were to be supplied locally after one year. Exports targets were set and engineering assistance, specifications and standards were to be issued by Singer to local suppliers.

Singer worked towards these goals and local content rose to 80 per cent by 1967. Although the domestic market absorbed most of the initial output, exports rose steadily. By 1975 Singer exported around 86 per cent of its production. In the early stages, Singer despatched experts in accounting, production management and engineering. It established training courses for local parts suppliers and worked with vocational high schools to train lathe turners and other technicians. The company needed to ensure that material input quality was up to a high export standard and that costs were minimized. It therefore happily supplied quality control training, product specifications, measuring equipment and in-plant training facilities for local toolmakers.

Taiwan's sewing machine exports grew from US$0.2 million in 1964 to around US$70 million in 1979. Total industrial output rose from 91,000 units in 1964 to 2,076,000 in 1979. Singer played a major part in this growth by training up around 140 parts suppliers, which formed the initial backward linkage effect. Singer was a role model for others to follow and, in some cases, to compete with. Some local producers learned to export directly through trading companies into Western markets.

The improved supply infrastructure led to the forward linkage effect, as four major Japanese sewing machine makers entered the market during the 1970s. The four companies, Taiwan Janome Sewing Machine Co., the Taibo Machine Co., Brother and Ishin Sewing Machine Co. (part of Toyota Group) benefited from the low-cost local suppliers and the marketing and informa-tion channels established earlier.[28] Some of the larger local firms (e.g. Lihtzer) began to outpace Singer in some segments of the market. As a result of the chain reaction initiated by Singer, in 1978 Taiwan overtook Japan as the largest supplier of sewing machines.

Athletic Shoes

The case of athletic shoes illustrates a new feature of clustering: the growing dependence of Western manufacturers on latecomer firms. During the 1980s the US firm, Nike, became a world leader in athletic shoes.[29] Nike's sales grew from around US$0.4 billion in 1984 to US$3.4 billion in 1992, while direct employment reached 7,800 workers worldwide. Around 600 of Nike's employees worked in East Asia to assist sub-contractors to meet quality and delivery targets. In 1992 most new shoe designs (around 100 per annum) and marketing was carried out in Oregon in the US.

By 1992, Nike's shoes were made almost exclusively by East Asian sub-contractors who employed around 75,000 workers, of which 80 per cent produced shoes, the balance apparel. Half of Nike's shoes were made by six large firms, three located in Taiwan and three in South Korea. The rest were produced by smaller sub-contractor factories. Four of the six had worked with Nike for at least 16 years. In Taiwan Feng Tay Co. produced 13 million shoes in 1992, Bao Cheng made 9 million and ADI Corporation, 7.2 million.

By the early 1990s much of Taiwan's shoe production had been relocated to lower-cost China (mainly Fujian Province), Thailand and Vietnam. In 1982 and 1983 Nike had a disastrous experience in China. It invested directly but made financial losses and lost market share to Reebok. Nike thereafter devised a strategy of using Taiwanese sub-contractors to operate Chinese factories, a pattern increasingly seen in electronics, bicycles and other sectors.

Purchasing by Nike, Reebok, L.A. Gear and others gave rise to a large cluster of sub-contractors in Taiwan as hundreds of SMEs surged into the market. Shoe production created a demand for leather processing, speciality chemicals, packaging systems, rubber moulding and other backward-linkage industries. In 1988, 1,245 firms produced shoe products in Taiwan. With relocation to lower-cost sites this number fell to 745 in 1992.

Firms learned many technical skills under sub-contracting arrangements. During the mid-1970s, making athletic shoes was a relatively unsophisticated activity, involving hand-stitched leather uppers and a simple moulding process.

However, by the early 1990s, production had become a complex, multi-stage operation involving synthetic materials, pigments, chemical agents, production automation and just-in-time inventory systems. By that time much of the manufacturing technology had transferred to Taiwan and South Korea.

Typically, from Nike's headquarters in the US, blueprints for new shoes would be relayed directly to Taiwan by fax or satellite. In Taiwan, local engineers would evaluate processes and manufacturing locations, taking into account design features, cost, quality and delivery requirements. Using computer-aided design technology, blueprints would then be turned into prototype shoes ready for production.

Through the 1980s, Taiwan's latecomers ensured the transfer of an increasingly complex technology which led, ultimately, to the reliance of many Western companies on East Asian manufacturers. Growing export opportunties gave rise to clusters of SMEs, some of which grew to become large firms. The sub-contracting system in footwear paralleled the OEM system in electronics: both systems provided latecomer firms with a means to overcome their market and technology disadvantages; both allowed for complex production methods to be learned by latecomers. As with electronics and sewing machines, the clustering effect in shoes led to more foreign investment and the birth of new supporting industries.

The Case of Bicycles

The final case, bicycles, illustrates how an entire sector, including manufacturing processes, design skills and brand names, was transferred from the US to Taiwan in a little more than a decade. This case emphasizes the rapid learning abilities of some latecomer firms and highlights the aggressive strategies used to gain US market share and to improve on foreign technology. As in the case of electronics, Western buyers and manufacturers provided the technology transfer channel, presenting the latecomers with the demanding users needed to stretch their technological competences.

During the 1980s, US sales of bicycles grew rapidly, reaching some 12.6 million in 1987. At that time around 7.4 million (59 per cent) were imported. US bicycle distribution was conducted through a complex web of mass merchandise chains, discount stores and large numbers of independent dealers, making entry difficult for new suppliers.

During the late 1970s, Taiwanese producers began to undercut Japanese manufacturers in low-end bicycles and several US buyers entered Taiwan to benefit from low-cost production. As Egan and Mody (1992 pp. 321–34) show, the links between US buyers and Taiwanese suppliers often endured for ten years or more. Through repeated transactions, local producers were able to learn about the US market and the capabilities of US bicycle makers.

As with the OEM system in electronics, US producers bought Taiwanese bikes in bulk, labelled them with their own brand names and sold them through their own distribution channels. US firms cooperated with local suppliers to ensure production quality and efficiency, leading on to collaboration in product design and development. In cases where trading companies acted as intermediaries, engineers from US firms (usually the final buyers) would be despatched to transfer product specifications and resolve quality problems. These direct connections enabled many local suppliers to dispense with the middlemen as they grew in size.

US buyers helped latecomer firms to understand changing American consumer tastes and the complexity of US distribution channels. The buyers sent experts to train local workers and technicians. They organized visits to US plants for training in production methods, paint finishing, packing materials and so on. As learning progressed, buyers would frequently insist that the Taiwanese firm not only manufactured the goods but also supplied the detailed product and process specifications.[30]

The history of the Schwinn Bicycle Company graphically illustrates the strategies used by latecomers to transfer the bicycle industry from the US to Taiwan. Schwinn was one of the largest US importers and distributors of bicycles, selling around one million bicycles per annum in the late 1980s. Founded in 1895, the company's Chicago home became the world's centre for bicycle production, boasting a cluster of around 90 manufacturers.

By 1992 Schwinn was the last remaining major bicycle maker in Chicago. Its US share had eroded from over 25 per cent of unit sales in the 1960s to 7 per cent in the late 1980s, falling to just 5 per cent in 1992. In 1992 Schwinn sold around 500,000 to 600,000 bicycles, mostly made by sub-contractors. Revenues were around US$200 million, and falling. After sustaining heavy debts, in 1992 Schwinn filed for bankruptcy.[31]

By contrast, in Taiwan the bicycle industry was thriving. One of Schwinn's main suppliers, the Giant Manufacturing Company, became the world's largest bicycle exporter in 1992 (in value terms). Giant began selling under its own brand name outside Taiwan in 1986 and by 1990 had become the leading exporter of branded bicycles to the US, Europe and Japan. In 1992 it sold around 1.35 million units and reached revenues of approximately US$230 million per annum.

This reversal of fortunes is partly explained by the technology and market strategies pursued by both companies. Schwinn began purchasing small quantities of Giant bikes in the late 1970s, selling them under its own brand name. In 1981 Schwinn management reacted to a strike by closing its main Chicago factory and transferring production capacity, engineers and equipment to Giant's, then small, factory in Taichung in Western Taiwan. Schwinn distributed the Taiwanese goods under its own brand label. In 1984 Giant sold around 700,000 bicycles through its US benefactor, accounting for about 70 per cent of Schwinn's sales.

Giant learned about and improved upon Schwinn's technology. For example, it led the market in carbon fibre bike frames, while Schwinn failed to introduce new innovations. During the 1980s, mountain bikes, which took over from racers, accounted for around 60 per cent of sales worldwide. Schwinn, which had led in ten-speed racers, fell behind with the transition to mountain bikes. On equivalent models Giant undercut Schwinn on price by 10 to 15 per cent.

With its growing knowledge of the US market and bicycle technology, Giant then launched its own brand, first in Europe (in 1986) and then in the US (in 1987). Schwinn's strategic response to Giant led directly to the blossoming of another dynamic latecomer. Schwinn reacted by extending its alliance with the Shenzhen China Bicycle Company (CBC) in China. CBC was formed in 1984 between Schwinn, a Hong Kong bicycle maker (Hong Kong Link Bicycles) and a Chinese partner, Shenzhen Light Industry Co. Link had produced low-end bicycles since 1969. In 1985 CBC began making low-cost, commodity bicycles for sale to Sears and other US mass merchandisers. However, in its alliance with Schwinn, CBC, like Giant, rapidly learned the techniques needed to produce mainstream and high-end, speciality bikes. At the same time, the alliance improved CBC's credibility with dealers and raised the quality of CBC's production standards. The Hong Kong engineers working for Link Bicycles acted as a conduit for technology transfer from Schwinn to CBC.

In 1990, under opposition from Schwinn's management, CBC (like Giant before), introduced its own brand business in the US by acquiring a medium sized US bicycle importer with a well-known brand name, Diamond Back, for US$17 million. Through the importer's distribution channels, CBC was able to compete directly with Schwinn. In 1992 CBC became the largest bicycle producer worldwide, measured in terms of unit output (roughly 1.8 million units per annum). Around 30 per cent of CBC's output was sold under the Diamond Back brand name.

Benefiting from low-cost production in China as well as innovative new products, CBC, Giant and other latecomers successfully gained market share from Schwinn and other US manufacturers. Schwinn's internal production fell to around 10,000 bikes per annum in 1992. Cumulative debts then forced Schwinn to file for bankruptcy.

A final twist to the bicycle case illuminates another important East Asian dynamic, namely the rapid growth of consumer demand in the region. In the early 1990s, China became the fastest-growing market for mountain bikes and, naturally, CBC and Giant took a lead in exploiting this new opportunity. In 1993 CBC sold an estimated 30 per cent of its bikes on the Chinese Mainland, up from ten per cent in 1991. To meet Chinese demand, CBC opened the world's largest bicycle plant in 1993 in the north of Shenzhen. Giant also began constructing very large new plants to supply the Chinese market. Needless to say, by this time most US manufacturers stood little chance of competing in China.

To sum up, the case of bicycles shows how an entire industrial cluster was transferred from the West to the East in little more than ten years. The latecomers acquired and improved upon existing technology, innovating in new ways. Their strategies enabled them to exploit overseas market opportunities and overtake the market leaders. As in the cases of electronics, sewing machines and footwear, Western companies supplied the necessary technology. By focusing their efforts on customer demands, the latecomers overtook their previous teachers.

In contrast with the electronics case, the bicycle producers made the full transition from OEM latecomer to ODM to OBM illustrated in Table 5.6. In electronics, the technology is more complex, the innovation frontier moves quickly and many US leaders learned to use the OEM system to their advantage, controlling key technologies and using patent and copyright protection to protect their position. By contrast, in bicycles US leaders lacked innovative and creative responses.[32] Their failure to manage the latecomer challenge led to industrial decline and the departure of many US firms from the industry. Whether or not the latecomers make the full transition to OBM in electronics will depend largely on the strategies and counter-strategies of both leaders and latecomers.

## 5.12	BEHIND THE FRONTIER INNOVATION

The case of Taiwan illustrates how so many East Asian latecomer firms were able to overcome the disadvantages of small scale. The scale-intensive approach of the South Korean *chaebol* contrasted sharply with the actions of SMEs in Taiwan. Taiwan's success in electronics relied on the speed, skill and agility of hundreds of local entrepreneurs, rather than the scale and financial power of the *chaebol*.

Taiwan's electronics firms followed a variety of strategies within a pluralistic and dispersed industrial structure. The fast-moving SMEs innovated across niche markets, whereas Tatung and other major manufacturing groups tended to follow a process-driven path similar to the *chaebol* but on a much reduced scale. Although local firms overtook the TNCs during the 1980s and 1990s, in terms of domestic production, foreign firms and locally-based joint ventures remained central to Taiwan's technological progress.

Even within the SME sector there was strategic variety. Some remained tiny specialist firms, dealing with mixtures of hardware, software and services.

Others, like ACER, grew to become medium-sized, international corporations. Although most firms relied on their own entrepreneurial ingenuity, some, especially in semiconductors, received direct support from the government.

Much of this chapter was devoted to explaining how innovation clusters were formed behind the technology frontier, not only in electronics but also in other fast-growing export industries. A model was put forward to show how the collective dynamism of latecomer companies attracted more TNCs and foreign buyers. Eventually the interaction of backward and forward linkages culminated in the large, innovative clusters of highly competitive firms in bicycles, footwear, computers and other important export markets. By the early 1990s many leading TNCs were dependent on Taiwan's manufacturing clusters, relying on them for their manufacturing and engineering services, components and finished goods.

Taiwan provided further insights into the difficulties of competing in own-brand products. Although many recent latecomers entered with state-of-the-art innovation capabilities, most were constrained to OEM and ODM, finding it very difficult to break through to OBM. On the positive side, firms grew rapidly under OEM/ODM, learning new technological and marketing skills and innovating with new designs. As an organizational innovation, the OEM/ODM system provided the technology and market lifeline for thousands of latecomer companies.

Like their ROK counterparts, Taiwanese latecomers developed the skills needed to exploit foreign channels of technology, using export market demand as the focusing device for technology investments. Harsh export requirements forced costs down and kept quality high. Through time, they learned how to narrow the innovation gap between themselves and the international leaders. Despite continuing weaknesses and setbacks, the innovation distance already travelled by the latecomers suggests that they are well-positioned to meet the fast-changing demands of the electronics industry.

NOTES

1. For an analysis of the exporting propensity of the SMEs see Chou (1992).
2. This is also confirmed by the Hong Kong example in Chapter 7.
3. The importance of industrial clustering is explained by Porter (1990). Normally such industrial clusters only occur in the developed world among centres of leading innovators and demanding users. In East Asia they occurred behind the technology frontier set by market leaders.
4. In 1941 Japan controlled around 91 per cent of all paid up capital in private business (Schive 1990 p. 9).
5. These became the in-house marketing branches of the *chaebol*.
6. For semiconductors see Chang et al. (1993); more generally for government high-technol-

ogy policy see Chiang (1988 and 1990) and Gwynne (1993). Wade (1990) looks in detail at government support policies. Few, if any, studies try and assess the failures and costs of direct interventions in the economy.

7. Taiwanese firms are famed for skilfully negotiating the government bureaucracy. What is sometimes called the parallel economy, built on a black market and underground financial system, may have comprised up to 40 per cent of the entire economy in 1993, according to government statistics (*The Economist* 6 November 1993, p. 91).

8. The term OEM/ODM is used because, in practise, OEM and ODM overlap considerably.

9. These restrictions were eased after the 1950s to encourage FDI, joint ventures and technical agreements.

10. In 1966, the government published a plan to establish Taiwan as an electronics industry centre, again mainly to attract investment (Wade 1990 p. 94). Overt government technology strategies came later, after the establishment of a strong, indigenous domestic sector.

11. The first local industries to generate significant exports for Taiwan were textiles and processed foods, built up during the Japanese colonial period. Local companies began by using their networks of overseas Chinese in Asia to sell their products. To expand further, they forged connections with Japanese trading companies and large US buyers in other growth industries. Unfamiliar and costly tasks such as shipping, trade finance and marketing were usually dealt with by the trading companies (Dahlman and Sananikone 1990 p. 44).

12. Taiwan became the second largest procurement centre for IBM in the 1970s. Between 1966 and 1987 it purchased US$1.3 billion worth of computer goods (Chaponniere and Fouquin 1989 p. 44).

13. By 1988 50 or so ASIC design firms operated in Taiwan, several of which later began manufacturing semiconductors (interviews, 1989). In 1978 ERSO of ITRI began Taiwan's first integrated circuit fabrication line (around 7 micron line width) by licensing in CMOS technology from RCA (a patent holder). ERSO sent engineers to RCA in the US for training in both design and manufacture. Engineers from RCA then helped to set up the pilot production line at ERSO. After two years of working on the process, around 150 ERSO employees formed a start-up company, UMC, in 1980. With ERSO's process technology, UMC supplied large quantities of semiconductors for calculators, watches, telephone handsets and other fast-growing markets. Roughly 35 per cent of the initial investment capital was government sourced. By 1991 UMC was a public company with around 1,680 employees and sales of US$180 million (interview with UMC, 1993). As well as UMC, ERSO span off the Electronics Testing Centre (in 1982), The Taiwan Semiconductor Manufacturing Company (in 1987), the Taiwan Mask Corporation (in 1988) and EMMT System (in 1989).

14. The success of ITRI/ERSO as an industry incubator cannot be dealt with in detail here. Hobday (1993) provides an assessment of its strengths and weaknesses.

15. Section 5.11 analyses how *groups* of firms formed innovation clusters.

16. Many of the strongest Japanese–Korean linkages were in consumer electronics and DRAMs, where Japan dominated.

17. The existence of ODM signified a new organizational innovation as well as a deepening of design capabilities.

18. Firms in this category include those spun off from ERSO/ITRI and those benefiting from the Hsinchu Science-Based Industrial Park facilities.

19. Because of the importance of clustering, the progress of traditional SMEs is dealt with in detail in Section 5.11 below.

20. See Schive (1990) for a thorough analysis of the importance of FDI to Taiwan's development.

21. Data on FDI are from Schive 1990 (pp. 3–19), unless otherwise stated.

22. This pattern of taking root occurred in Singapore and Hong Kong. See the case of Motorola (Section 7.10) which promoted a pioneering ethnic Chinese manager to become senior vice president in charge of East Asia.

23. Spending on this type of activity reached around US$17.6 million in 1984, roughly 3 per cent of annual turnover.

24. Evidence for this section is from interviews in Taiwan in 1992 with ACER, Johnstone (1989 pp. 51–2), *Business Week* (June 28 1993 p. 38), the trade press especially *Electronics* (22 November 1993 p. 14), *Computrade International* (August 15 1993 p. 72), and Chaponniere and Fouquin (1989 p. 61).

25. Although the government did not provide funding it did allow tax incentives (20 per cent of initial investment was tax deductible) and provided R&D grants in some areas (e.g. multimedia technology).

26. In developed countries, industrial clustering is assumed to involve close proximity with users as well as the chief international sources of technology. The issue of *how* such clusters are created in latecomer economies lacking such conditions is not yet dealt with in the literature.

27. These figures differ from Levy's (1988) cited earlier. The source of III (1988) figures was a government commissioned survey conducted in Taipei. This probably understated the total number of computer makers and software design houses.

28. Vernon (1960) describes how such externalities led to industrial clustering in New York in the 1950s. The Taiwanese case is more striking, given that the economy began with many industrial disadvantages and few leading-edge users.

29. Data on Nike are from Clifford (1992a) pp. 57–9.

30. This mirrored the transition from OEM to ODM in electronics.

31. The following details are from *Forbes* (December 1992 p. 90) unless otherwise stated.

32. Freeman (1974 p. 256) argued that 'not to innovate is to die. Some firms actually do elect to die...'. It may be that the US bicycle producers allowed themselves to exit, by choosing not to make the innovative effort required to compete and survive.

6. Singapore: a test case of leapfrogging

6.1 LEARNING WITHIN TNCS

This chapter investigates the patterns of technological learning within Singapore's electronics industry, showing how firms, mostly TNCs, accumulated technology. Singapore has been a remarkably successful exporter of electronics. With a population of about three million in 1991 it exported around S$26 billion (US$15 billion) worth of electronics, double the export value of Hong Kong (US$7.5 billion), more than Taiwan (US$12.2 billion) and approaching the level of South Korea (around US$19 billion in 1990).[1] Electronics accounted for about 40 per cent of Singapore's total manufacturing output in 1992 (and around 40 per cent of its exports).

In contrast with the other three dragons, local industry played a very minor part in Singapore's electronics industry. Instead, development relied largely upon foreign TNCs. The TNCs enabled Singapore to export and to systematically acquire technology. The case of Singapore shows that with appropriate government policies, TNCs can be persuaded to train local employees and transfer technology to their subsidiaries, an important finding for developing countries wishing to harness TNCs for national development.

This chapter has two purposes. First, it explains why Singapore took the TNC route to development in electronics and why the TNCs chose to locate in Singapore. It examines the patterns of technological learning within firms and shows how the subsidiaries closed, in part, the technology gap with the TNC parent plants located abroad.

Second, it uses the evidence to examine the so-called leapfrogging argument, an idea popular among many in the technology and policy analysis field. The leapfrogging idea holds that with the diffusion of information and electronics technology some developing countries may be able to leapfrog older vintages of technology, bypass heavy investments in previous technology systems and catch up with advanced countries (see Section 6.2). According to some observers, Singapore developed quickly using information technology (Gilbert 1990; Sisodia 1992). It is therefore an interesting test case of leapfrogging.

Section 6.3 briefly traces Singapore's historical developments, highlighting Singapore's technological and educational policy. Sections 6.4 and 6.5

analyse the contribution of electronics to the economy, showing the structure and specialization of the electronics industry. The rest of the chapter examines the mechanisms of corporate learning, looking at TNCs (and two local firms) in each of the four leading electronics sectors (semiconductors, disk drives, consumer electronics and computing).[2]

6.2 THE LEAPFROGGING ARGUMENT

The diffusion of the information technology paradigm and the rapid growth of the electronics industry have led some observers to believe that some developing countries may be able to leapfrog older vintages of technology and begin to catch up with advanced economies.[3] According to this view, some industrializing economies are less hampered by commitments to previous generations of technology. Developing nations with adequate levels of skills and infrastructure (absorptive capacity) may benefit from the windows of opportunity provided by the new paradigm, especially at the early stages of diffusion when barriers to entry are low and markets are in a state of upheaval. Leapfrogging is consistent with Schumpeter's notion of gales of creative destruction, where radical new innovations give rise to the destruction of industries based upon older technologies (Schumpeter 1950).

In the area of telecommunications infrastructure, research shows that many developing countries have indeed adopted digital, electronics-based systems more rapidly than advanced countries.[4] Less committed to older vintages of technology, they jumped directly to advanced electronics systems. In fact, they had little choice in the matter. Digital systems are generally less expensive, more robust and flexible than older technologies. Most major suppliers therefore stopped making older systems and concentrated on digital equipment.

However, the position with respect to firms and industrial development is less clear. There has been little empirical study of whether or not firms in developing countries can leapfrog – or indeed what the term actually means in practice. As Archambault (1992) argues, the leapfrogging thesis is based on theory rather than empirical research.[5] Furthermore, as Chapter 3 indicates, many studies show that firms acquire technology through a costly, difficult and incremental learning process. On the face of it, the notion of learning contradicts the idea of leapfrogging.

This chapter takes the opportunity to use the case of Singapore to assess the leapfrogging hypothesis. The economy has made a remarkable transition from developing to developed country status and has developed a dynamic electronics and information technology infrastructure. By looking at technological progress at the firm and industrial level it is possible to see whether or not Singapore's experience is consistent with the leapfrogging argument.

In several respects Singapore is a special case among developing countries, not only due to its small population and regional trading status, but also because the government strongly promoted education and developed a modern telecommunications and electronics infrastructure.[6] The country appears to have quickly developed the absorptive capacity required for leapfrogging to take place. Indeed, electronics has grown rapidly to become the largest industrial sector and, according to some observers, information technology was instrumental in promoting Singapore's growth (Sisodia 1992; National Computer Board 1992).

While one cannot generalize from a single case study, the Singapore case is indicative of whether leapfrogging can be expected to occur. Singapore lacked an extensive industrial base in the 1960s. It rapidly developed an absorptive capacity and was able to access technology through foreign firms. If industrial leapfrogging did not occur in Singapore, it is very unlikely to occur in other developing countries.

6.3   SINGAPORE'S ECONOMIC DEVELOPMENT

Historical Context and Government Policies

Singapore, a regional economic hub in Southeast Asia, has the second largest port in the world, providing access to Malaysia, Indonesia and other neighbouring economies. Located at the tip of the Malay Peninsula, it is also on the international shipping route between Japan and the Middle East. Founded in 1819 by Sir Stamford Raffles, the Island initially served as a trans-shipment point for British trade with East Asia.

During the 19th century Singapore became a regional entrepôt as the British extended their control over the Malay Peninsula (Haggard 1990 pp. 101–2). In order to develop the rubber and tin industries in the region, the British worked through a network of Chinese intermediaries. A complex infrastructure of transport, financial and commercial services, insurance and communications gradually developed to facilitate Singapore's entrepôt trade.

While some trade-related industries developed (e.g. tin and rubber processing, ship repair and maintenance, and some local consumer goods production), by the 1950s the local business community consisted largely of small merchants and financiers (Vogel 1991 p. 77). Family firms also operated in food processing, real estate, printing and other trade-related services. In 1960 the manufacturing sector was still very small and fragmented, accounting for just 12 per cent of GNP. Around 94 per cent of Singapore's exports were re-exports (Haggard 1990 pp. 101–2). Much of the Chinese-controlled local industry consisted of small family workshops connected to the retail trade.

Singapore began its drive for industrialization in the early 1960s, responding to problems of high unemployment and poor housing. During the latter half of the 1950s both trade and economic growth were sluggish, while political frictions with Malaysia and Indonesia brought additional concerns over Singapore's dependence on trade. Growth through industrialization was seen as the solution but, Lee Kuan-yew, Singapore's long-standing leader, believed that Singapore's entrepreneurs lacked the finance and management skills needed to bring about industrialization.[7]

Following self-government status in 1959, Singapore embarked upon an import-substitution strategy. The government believed that the political merger between Malaya, Sabah, Sarawak and Singapore in 1963 would constitute an internal market large enough to support the growth of Singapore's domestic infant industries (Yue 1985 p. 260). However, after political separation from Malaysia in 1965 the common market proposal was aborted. Further uncertainty was added when, in 1966, the British Government announced its intention of withdrawing its military bases within five years. These employed around 40,000 workers directly or indirectly (Haggard 1990 p. 110). In 1967, two years after independence, the import-substitution strategy was officially abandoned for export-promotion (Krause 1987 p. 60).

After Lee Kuan-yew's party (the Peoples Action Party) consolidated power and limited political opposition, with a small group of close economic advisers, including Goh Keng-swee (the Minister of Finance), Lee took direct responsibility for economic policy. According to Vogel (1991 p. 77) Singapore developed an administrative state, with government agencies taking a leading role in business through both direct and indirect means. Other writers refer to an entrepreneurial or corporate state (Sisodia 1992 and Yuan and Low 1990 respectively). State-owned or controlled enterprises were set up in oil exploration, petroleum refining, petrochemicals, defence, shipbuilding and airline travel. In the high-technology export sector, including electronics, the state left most decisions to private firms, providing incentives, manpower training and infrastructure to attract foreign companies to the economy.

The government decided that to generate employment and increase national income it was essential to promote industrialization and diversify away from Singapore's traditional entrepôt business. It believed that the most effective way of building up the electronics industry was to attract foreign TNCs. The policy makers judged that the domestic entrepreneurial base was too weak and unskilled. To attempt to transform them would have been too risky, slow and uncertain (Yue 1985 p. 260).

To induce the TNCs into Singapore, the state allowed a level of foreign control which would not have been acceptable in either South Korea or Taiwan. In 1967 and 1968 the government introduced restrictive labour legislation, new investment incentives, and educational measures for supplying more technically trained workers. The Economic Expansion Act of 1967 accelerated depreciation and allowed duty free imports of equipment and inputs for industry.

The Economic Development Board (EDB) established in 1961, became responsible for Singapore's industrial development. Partly as a result of the policy measures, foreign investment and manufacturing increased rapidly. The growth in world trade and international investment led to rising exports and wages and, as early as 1970, full employment was reached (Krause 1987 p. 60).

TNC Investment Decisions

During the 1960s and 1970s, TNCs came to Singapore in response to political stability, the geographical location and the efficient and constantly improving transportation and communications infrastructures. The freedom granted to foreign firms, administrative support from government and macroeconomic stability were also important. Several investment incentive schemes operated, although few firms in the electronics sector invested because of government incentives alone.[8]

Some firms were initially attracted by the low costs of labour and overheads. Philips of the Netherlands located in Singapore in 1951 and gradually expanded its consumer appliance business. Jurong Industrial Park, which was set up in 1961, attracted more TNCs. US firms such as GE and HP set up operations in Singapore, as did NEC and Fujitsu of Japan. In semiconductors, TI of the US and SGS of Italy both set up assembly and testing facilities in 1969. Although many of the TNCs operated in electronics and electrical machinery, according to EDB officials, the electronics industry was not especially chosen or targeted by the government (interviews, August 1992). All industries which could create employment and exports were welcomed. Few, if any, would have been rejected.

Several times the EDB attempted to raise the standard of technological investments, but then reversed its policy due to economic circumstances. In the early 1970s it began to try and attract more skill-intensive, higher value-added export industries. But in 1974 and 1975 the oil crisis and the ensuing recession in the industrialized countries led to a slowdown in Singapore's growth (to around 5 per cent). The government responded with a temporary postponement of the higher technology policy, a new policy of wage restraint and a return to encouraging labour-intensive investments. In 1979 efforts to upgrade the industrial structure were re-intensified and the National Wages Council recommended large increases in wages in order to accelerate the trend towards technology-intensive manufacturing (Krause 1987 p. 61). However, in the 1985 to 1986 recession (caused by a decline in demand for

Singapore's oil and marine related services and the slowdown in the US electronics business), the policy was reversed again. In response to Singapore's crisis, the government called for a severe wage restraint and relaxed its high-technology policy (Krause 1987 pp. 10–18).

Despite cyclical recessions and booms, by the early 1990s Singapore had built a modern industrial sector and an advanced telecommunications and transportation infrastructure. The island had become one of the most popular headquarters locations in Southeast Asia for large and small TNCs, many investing in high-technology manufacturing activities.

By 1990 more than 3,000 TNCs (including 600 large firms) from the US, Japan and Europe operated in Singapore. Cumulative foreign manufacturing investment in Singapore increased from around S$7 billion in 1980 to S$13 billion in 1985 to nearly S$24 billion in 1990.[9]

In 1991 total manufacturing investment amounted to US$2.5 billion of which nearly US$1.2 billion (48 per cent) was in electronics. In electronics the US invested US$582 million in 1991, followed by Japan with US$391 million and Europe with US$240 million. Other Asian NIEs accounted for US$6 million.[10] By 1992 more than 250 electronics firms together sold in the region of S$31 billion (US$19.4 billion) in electronics.

Policies for Education and Training

In order to promote manufacturing the government set up policies for education, training and skills development. Following the basic technical education provided by vocational institutes in the 1960s, during the 1970s the EDB organized apprenticeship schemes and government–industry training centres with Japan, France and Germany. By 1991 the EDB operated five training institutes with an enrolment of 2,500 students: the French–Singapore Institute, the German–Singapore Institute, the Japanese–Singapore Institute, the Precision Engineering Institute and the Philips–Government Training Centre.

The main function of the institutes was to provide engineering, technology and craft education for manufacturing industry. The institutes provided two and three year training courses in tool and die and precision machining, plastics technology, factory automation, mechatronics and industrial electronics. Many other programmes of re-training and continuous education operated to support manufacturing industry in the early1990s.

By 1991, the National University of Singapore, Nanyang Technological University, the polytechnics and the training institutes together supplied around 22,000 engineers and craftsmen per annum, an annual output of roughly 38 per 100,000 population, one of the highest levels worldwide per capita.[11] The availability of low-cost engineers, technicians and skilled workers was cited by major TNCs as a primary reason for locating in Singapore (see below).

Another instrument of training policy was the Skills Development Fund (SDF), set up to provide training grants of between 30 per cent and 70 per cent of total costs. During the early 1980s the SDF was financed by a levy on employers of around 4 per cent of workers' wages, matched by government contributions. As of 1992, the SDF was funded by a levy on employers of around 1 per cent on wages less than S$750 per month. Many firms received funds to finance training programmes.[12]

6.4   POLICIES FOR ELECTRONICS

Technology Institutes for Electronics and IT

In the early 1990s, the government set up institutes for software training, electronics engineering, advanced mechanical engineering and research. This was part of the government's renewed policy to encourage manufacturing into higher technology activities such as design and research and development (R&D) and thereby take root in Singapore. Technology support institutes for electronics and information technology in 1991 included the Institute of Manufacturing Technology, the Information Technology Institute, the Institute of Systems Science, the Institute of Microelectronics and the Magnetic Technology Institute (EDB 1992a).

At the time of the research the institutes had yet to have a major impact on industry. The government hoped that they would supply the high-level skilled manpower, engineering and research back-up to encourage more TNCs to further upgrade their local technological activities. The advisory boards of the institutes often included managers of TNCs who advised on specific projects and strategic direction.

The new institutes are invariably targeted at high-technology sectors. The Magnetics Technology Centre, for example, carried out engineering and research work for the disk drive industry. Its work included the design and control of actuators and motors, magnetic data coding and electromagnetic field computational techniques. The Institute of Microelectronics was set up in 1992 to support the semiconductor industry with silicon processing technology, component reliability, failure analysis, computer-aided circuit design and systems applications. One of its long-term aims was to enable the large chip producers to carry out complex developmental work locally.[13]

Weaknesses in Advanced Technology and Science

In contrast with its strong supply of technicians and production engineers, Singapore lags behind other countries in the supply of research engineers and scientists. In 1990 Singapore had around 28 research engineers and scientists per 10,000 workers (a total of 4 300), compared with 87 for Japan, 77 for the US, 44 for Switzerland, 43 for Taiwan and 33 for the Republic of Korea.[14] Analysis of government spending on R&D as a percentage of GDP also shows that Singapore lags behind many developed countries, as well as Taiwan and the Republic of Korea.[15] Singapore's lack of research engineers and scientists in part reflects the past priorities of industry. During the 1970s and 1980s industry did not require a large supply of high-technology researchers and scientists. As the company case studies below show, the main demand was for skilled and semi-skilled workers and production engineers.

Future Policy Plans

In the early 1990s, the government planned to encourage further high-technology manufacturing, especially in electronics. A large effort was being made to build up the science and technology infrastructure of the economy. Under a National Technology Plan laid out in 1991, the National Science and Technology Board aimed to spend S$2 billion (roughly US$1.2 billion in current prices) over a five year period to increase the number of researchers from 28 to 40 per 10,000 workers. By making this investment the government hoped to lift the economy onto a higher plane of research-intensive, higher technology activities during the 1990s.

The government also established a so-called Growth Triangle to shift labour-intensive production out to Malaysia and Indonesia, while retaining higher wage activities in Singapore. The Growth Triangle, first discussed in 1989, consisted of two adjoining low-cost regions within neighbouring Malaysia (Johor) and Indonesia (the Riau Islands). In 1990 agreement was reached to develop and manage a 500 hectare industrial estate (the Batam Industrial Park in Indonesia) 20 kilometres south of Singapore. Batam is around two-thirds the size of Singapore and, in 1993, wages were less than one-half of Singapore's (Clifford 1993 pp. 60–63). Batamindo is Batam's industrial complex, formed through a joint venture between Salim Group and Bimantara (two Indonesian business groups) and the Singapore Government.

By the end of 1992, 34 corporations had invested around US$119 million in Batam, including AT&T and Telemecanique (of France), Astra M.T. (of Indonesia), Sumitomo, Philips, Thomson and Bowater Packaging. The government-owned Singapore Technologies Industrial Corporation was reported to have allocated US$5 billion for infrastructural development in the Triangle.[16]

Not unexpectedly the Growth Triangle encountered some problems, not least the reluctance of the three governments to form an agency to manage the region. Economic links between Johor and Riau had failed to materialize as planned and the political arrangements were complex, with only informal agreements being endorsed at high political levels. In Malaysia there were worries that the Johor development might pull investments away from the northern part of the state. Although the Triangle accelerated the development of Riau in Batam, its overall future remained to be seen.[17]

6.5   ELECTRONICS AND THE ECONOMY

Singapore's economic achievements are well documented. GDP per capita rose from US$435 in 1960 to US$13,236 in 1991.[18] The structure of the economy changed significantly as a result of economic progress. Entrepôt trade declined from more than 60 per cent of total exports and roughly 20 per cent of GDP in 1965 to 35 per cent of total exports and 9 per cent of GDP in 1985 (Krause 1987 pp. 64–5). Manufacturing grew from around 18 per cent of GDP in 1960 to 27 per cent in 1991, accounting for more than 50 per cent of Singapore's foreign exchange earnings.

Within manufacturing, electronics was the largest sector in 1991, accounting for 34 per cent of gross manufacturing value-added (EDB 1992a p. 21). In 1992 electronics accounted for around 40 per cent of production and exports (*Electronics* 26 April 1993 p. 5). After electronics, the next largest sector was chemicals and chemical products.

Total electronics output rose from S$24.5 billion in 1989 to S$31 billion in 1992. The bulk of production was carried out by large TNCs for export. Exports increased from S$21.2 billion in 1989 to S$26 billion in 1991, while employment increased from 116,000 in 1989 to around 124,000 in 1991 (Table 6.1). Total value-added for the industry was estimated at S$7 billion in 1989 to S$7.8 billion in 1991 (roughly US$4.5 billion) (EDB 1992b p. 2).

Table 6.1 outlines Singapore's main electronics production for 1991. The largest product group, disk drives, accounted for S$7 billion worth of exports in 1991 (around 27 per cent of total electronics exports). Singapore became the world's largest producer of hard (i.e. Winchester) disk drives in the early 1990s, producing around 50 per cent of global output in 1991 and 1992. In 1992 around 18 million disk drives were manufactured, an increase of 28 per cent on the year before.[19] During the 1980s, Singapore attracted a large cluster of international producers, including Seagate and Conner of the US.

Exports of semiconductors, the second largest electronics sub-sector, amounted to around S$4.5 billion in 1991 (17 per cent of total exports). Industrial electronics overall (including disk drives, semiconductors, computers and printed circuit boards) represented 56 per cent of total electronics exports in 1991. During the 1980s Singapore made the transition from cheap-labour assembly to an advanced, automated manufacturing.

Table 6.1   *Electronics exports and employment by major product*
                categories (S$ billions)

Selected product lines	Exports			Employment
	1989	1990	1991	1991
Disk drives	5.5	7.2	7.0	25,000
Integrated circuits	3.7	3.8	4.5	16,000
Printed circuit boards	2.6	3.0	3.1	13,870
Computer systems	1.0	1.2	2.4	2,577
Audio equipment	2.1	2.0	2.0	n/a
Colour TVs	1.3	1.4	1.4	n/a
TV and audio sub-assemblies	1.5	1.5	1.4	n/a
Telecommunications	0.7	0.8	0.7	8,000
Product groups				
Industrial electronics	11.0	13.6	14.7	n/a
Consumer electronics	5.4	5.6	5.7	26,000
Components	4.8	4.9	5.7	n/a
Total	21.2	24.1	26.1	123,516

Source: Compiled from EDB (1992b).

The contribution of local firms to the electronics industry was very small. In 1991 Singaporean-owned firms accounted for around 16 per cent (approximately S$472.9 million) of total manufacturing investment, much less for electronics investment. Some local firms supplied the TNCs, but most of their materials, components and capital goods requirements were imported or made locally by other TNCs. Some domestic firms (e.g. computer makers) provided additional capacity for TNCs through OEM and sub-contracting arrangements. A few local companies supplied plant and engineering support for large firms.

A small number of local firms grew to become medium-sized TNCs in their own right, including Singapore Technologies Industrial Corporation and Wearnes Technology, the two largest firms. Other major firms were PCI Inc., Eltech Electronics and Goh Electronics. Combined electronics sales of the top five firms amounted to around US$0.7 billion in 1989, a tiny proportion of total output.[20] Local firms produced PCs, Winchester disk drives, printed circuit boards and data communications equipment.

Although local computer manufacturing tended to be carried out under OEM arrangements, a small number of local firms developed their own brand names (e.g. Wearnes and Creative Technology), controlled their own distribution and carried out R&D. Indeed, Creative's Soundblaster technology became the *de facto* standard for PC sound cards, eclipsing US versions. Potentially even more significant was its entry into the digital video board market with its Videoblaster technology. In the US Creative also played a leading part in establishing an industry standard for digital compression technology for PCs, showing how this particular company was able to compete at the technology frontier. However, in comparison with Hong Kong, Taiwan and the Republic of Korea, the contribution of local firms to electronics output overall was very small in Singapore.

Table 6.2   Electronics study sample: employment, sales and origin of companies

Sector	Firm	Start Date	Origin[a]	Employment	Sales 1990 (S$m)[b]
Semiconductors	TI	1969	US	2,000	1,123
	SGS-Thomson	1969	I/F	1,000	1,194[b]
	NEC	1976	J	700	330
	TECH	1990	JV	420	71
	Chartered	1989	S	400	450[b]
Disk drives	Seagate	1982	US	12,000	4,833
	Conner	1987	US	4,300	2,000
Consumer electronics	Philips	1951	N	6,100	2,000[b]
	AT&T/ACP	1986	US	4,000	n/a
Computers/ professional	Wearnes[c]	1983	S	510	128

Notes:
[a]   Codes: US=United States; I/F=Italy/France; J=Japan; N=Netherlands; S=Singapore; JV = Joint venture between Singapore's EDB (26 per cent) and US and Japanese firms.
[b]   Denotes 1991 data (exchange rate for 1990 US$1=S$1.81; for 1991 US$1=S$1.73.
[c]   Includes Wearnes Hollingsworth and Wearnes Automation.

Sources: Interviews, company annual reports and *Singapore Electronics Manufacturers Directory* (various issues).

6.6 CORPORATE TECHNOLOGY STRATEGIES

As in the other three dragons, open-ended questionnaires were used to identify the key categories of learning and to describe how these changed through
time. In Singapore, the firm sample was structured to include: (a) leading
TNCs; (b) local and joint venture firms; (c) large, medium and small companies; (d) early and late entrants; (e) firms from four of the major product
groups; (f) firms from Europe, the US and Japan.

As Table 6.2 shows, total sales of the firms interviewed amounted to
around S$12.1 billion in 1990. Estimating additional sales of roughly $S2
billion for AT&T, this represents just over one-third of Singapore's electronics output.

6.7 SEMICONDUCTORS

Sector Overview

In 1991, the chip industry exported around S$4.5 billion, making it the second
largest electronics sub-sector in Singapore. Among the leading TNCs were TI,
National, AMD, NEC, AT&T Microelectronics, Siemens, SGS-Thomson,
Fujitsu, Matsushita, TECH Semiconductor, Chartered Semiconductor, Unitrode
and Linear Technology. Semiconductor activities began with assembly and
testing. In 1992 around 5 per cent of the world's chip testing was still carried
out in Singapore. During the 1980s, designs for semi-customized chips and
wafer fabrication were introduced selectively. Companies also introduced
regional management, marketing and distribution for the Pacific Asia region.

In 1969 TI became the first chip maker to begin production in Singapore.
TI was followed by SGS of Italy (now SGS-Thomson of Italy and France),
also in 1969. NEC Singapore was incorporated in 1976.[21] The three TNCs
chose Singapore because of its efficient air and sea ports, the relatively low
cost and high skills of the labour force and government assistance of various
kinds. In each case, operations began with labour-intensive production, primarily the assembly and testing of mature products such as discrete semiconductors, linear and bipolar integrated circuits.

In each of the TNC cases, the choice of location was determined by: (a)
corporate strategies towards internationalization and capacity expansion; (b)
the way semiconductor technology allows firms to relocate parts of the production process overseas; and (c) host country attractions (e.g. the infrastructural advantages of Singapore over competing locations). The TNCs
found Singapore a congenial and efficient location in which to set up mature,
labour-intensive operations and then to expand and upgrade.

Upgrading of TNC Subsidiaries

Over time, the three TNCs introduced more complex products requiring greater engineering support, sophisticated processes, personnel training, some automation and more intensive quality control. NEC, for instance, went through a series of product transitions, beginning with transistors in 1977 and linear integrated circuits in 1979. By 1986 the factory assembled 256k DRAMs and, by 1991, the 1 and 4 megabit DRAM. TI followed a similar path. Initially, wafer fabrication took place outside Singapore in the TNC parent plants overseas.

With each product transition, testing and assembly became more complex and automated. In 1980 the ratio of test operators to machines was roughly 1:1. By 1991 the ratio was roughly 1:4. Several tasks remained labour-intensive (e.g. loading and unloading), requiring large numbers of employees.[22] Later introductions, (e.g. the 4 megabit DRAM), required continuous engineering development work (e.g. for failure analysis and advanced packaging). This led firms to cooperate with local university engineering departments and, more recently, the government's Institute for Microelectronics.

Introduction of Wafer Fabrication by the TNCs

Wafer fabrication is the core manufacturing technology activity in semiconductors. As such, it requires investments in skills (especially engineering), plant, machinery, technology and management: a step jump from testing and assembly. In 1984 SGS-Thomson became the first chip maker to set up wafer fabrication in Singapore (also the first in Pacific Asia outside of Japan). After 1984 the company set up three new modules, expanding the wafer capacity of its Singapore plant. Following SGS-Thomson, in 1990, TI became a partner in a sub-micron (i.e. leading edge) joint venture called TECH Singapore, to fabricate 4 and 16 megabit chips. TECH was a shared US$330 million venture between TI (26 per cent), the Singapore Economic Development Board (26 per cent), Canon (24 per cent) and HP (24 per cent).

The cases of TI and SGS-Thomson illustrate the technological upgrading of chip firms within Singapore. The TECH facility, scheduled to begin production in March 1993, aimed to supply regional users of DRAMs (including HP and Canon). Although NEC had no plans to introduce fabrication to Singapore, it had set up a design centre staffed with engineers.

Firms gave several reasons for fabricating in Singapore. First, the growing importance of the Pacific Asian market justified investments closer to customers (TI's sales to the region were around US$1.5 billion in 1992, roughly 30 per cent of its worldwide non-military turnover) (Clifford 1992 pp. 62–4). Second, the TNCs were able to achieve high levels of capacity utilization, due to Singapore's efficient infrastructure and the willingness of management and workers to work long overtime hours. Third, firms benefited from policies organized through the EDB. Fourth, the relatively low cost of production engineers in Singapore, compared with Europe, the US and Japan proved important. One company estimated that the entry level salary for an electronics engineer in Singapore was roughly 50 per cent of the cost of an American equivalent.

TNC Plant Expansion

By 1991, each of the three TNCs had expanded their plants considerably and set up regional and marketing headquarters. TI, for example, employed around 2,000 people, of which 1,000 were line workers and the balance engineers, technicians, marketing and other administrative staff. By 1991, SGS-Thomson's sales reached US$690 million, making the local plant one of SGS-Thomson's largest international operations. Each company had introduced design support facilities for local customers (e.g. ASIC design offices). SGS-Thomson began product design in 1979 and by 1992 employed around 30 full-time design engineers.

Most senior engineers and managers had been trained and promoted from within the Singapore operations. Two of the three TNCs carried out extensive computer-based training, developed by engineers within the local plants. According to the firms, local graduates were adequate for most engineering and management tasks. The need for additional technical back-up had led all three firms into arrangements with university engineering departments as well as the government's Institute for Microelectronics (IME). SGS-Thomson, for instance, conducted some product development work with the local universities and polytechnics and had donated equipment for use by the IME.

Plants were able to expand as a result of Singapore's continuous improvements in educational quality, the supply of engineers, and the improving transportation and telecommunications infrastructures. Government incentives were introduced to encourage TNCs to locate headquarter activities in Singapore. In the TI case, direct involvement by the government through its risk-sharing investment encouraged the firm to expand.

Technological Catching up by the TNC Subsidiaries

As new products were introduced, more mature products (e.g. bipolar TTL, ceramic and linear integrated circuits) were transferred to Malaysia and other low-cost sites. In terms of the product life cycle, Singapore progressed towards the newer (early stage) products, requiring a larger number of more complex production activities than before.

Absolute technological progress occurred as the Singapore subsidiaries graduated to more complex products and processes. TI's joint venture (TECH) and SGS-Thomson's wafer fabrication facilities (and HP's gallium arsenide fabrication facility and Chartered's manufacturing facilities) also indicate the *relative* catch-up with TNC parent firms in the developed economies. The gap between parents and subsidiaries narrowed in response to regional market growth, increasing labour costs, and improved absorptive capacity in Singapore.

However, the catch-up process was by no means complete. In 1992, little mainstream product design or R&D was carried out in Singapore. Highly advanced manufacturing tended to be introduced first to headquarter locations and only afterwards to Singapore. Product output in Singapore was advanced and mainstream, but not at the leading edge (e.g. 4 and 16 megabit DRAMs, rather than 64 megabit). TI, for instance, conducted most of its product design and R&D at the Dallas headquarters, while the local plant progressively learned more about failure analysis, packaging design and other near market-technology activities.

As the TNC subsidiaries advanced technologically, they increasingly integrated their activities within the Southeast Asian region. This is indicated by the growing number of forward and backward linkages among firms. SGS-Thomson, for instance, forged backward linkages with local suppliers of capital goods (e.g. Varian and Ultratech), substrates, gases, chemicals and other materials, some of which were based in Singapore, others nearby in the region. TI formed close forward links with customers such as Canon and Hewlett-Packard (HP). Sales to clients within Singapore increased. For instance, Seagate, the disk drive producer, became a major customer of SGS-Thomson. All three TNC subsidiaries established design centres and marketing headquarters to work more closely with regional clients. They also formed educational links with local universities.

A Local Semiconductor Company

Turning to the locally owned start-up, Chartered Semiconductor Manufacturing Pte. Ltd, this was the first Singaporean company to begin wafer fabrication. Chartered began in 1989 as a joint venture between the Singapore Technology Industrial Corporation (STIC), Sierra Semiconductor and National of the US. STIC was a part of Singapore Technology Holdings, a large state-owned sub-contractor to the Singapore military which employed roughly 12,000 people in 1991. Chartered, a producer of sophisticated chips (mainly mixed signal, analog and digital devices), began with start-up capital of roughly US$40 million, supplied by the Singapore Government (*Electronics Times* 19 November 1987 p. 48). Total investment increased to around US$123 million in 1991, by which time the company employed around 400 staff, of which 70 were engineers (*Electronic Business Asia* February 1991 p. 36).

Chartered jumped in at more or less world-level technology (1 and 1.2 micron process) with 20 or so domestic R&D staff. After several teething problems National pulled out of the operation in 1989 and Sierra departed in 1991. In 1991, a new president, formerly the head of the Taiwan Semiconductor Manufacturing Corporation, was hired. Under a revised strategy, the firm re-modelled itself on the Taiwanese company, fabricating semiconductors designed by other, usually US, firms.

Chartered learned by hiring, by licensing and by forming alliances with proven technology suppliers. The company was able to reshape its strategy due to financial backing from the government-owned STIC. A less well financed company could well have failed.

Implications for Learning and Leapfrogging

The case of Chartered shows how high-technology entry may take place once market and infrastructural conditions are suitable. As a latecomer, entry was facilitated by international technology suppliers and overseas experts. Chartered is an example of a large, well-financed, high-technology firm diversifying into a new business with government support. Whether it succeeds or not remains to be seen.

Among the TNC subsidiaries various forms of catch-up learning occurred over time. These ranged from production to innovation learning, from elementary to advanced learning, and from technological to non-technological learning. As learning by assembly and testing became more complex, tasks involved more in-house, formalized training. Learning from foreign engineers and managers was widespread and senior foreigners continued to oversee some operations, assisting in the installation of new capital equipment and so on. NEC, for instance, employed 29 senior Japanese staff in its plant of around 700 employees.

Leapfrogging is not a term which captures the cumulative nature of technological progress among the TNCs. Within firms, learning by experience and promotion occurred gradually. All firms recruited, trained and promoted local employees to senior positions in engineering and marketing, building up their technological and organizational capabilities. By 1992, the companies had acquired capabilities in product design, process adaptation, continuous engineering and selective R&D back-up.

Firms learned to market directly to customers nearby as the Pacific Asian market grew. Customers demanded further services, requiring further learning on the part of the TNCs: learning to design application-specific integrated circuits (ASICs), learning to market and learning to provide after-sales services.

In contrast with the leapfrogging hypothesis, firms progressed incrementally in response to market growth, local labour costs, the supply of engineers, and the improving infrastructure. Learning by imitation took place when firms followed others into the region, clustering in Singapore. Once one company had taken the plunge and proved successful in assembly, test and wafer fabrication, others followed. The local firm learned by hiring and by imitation. Recent evidence hinted at the start of learning between local input suppliers and the chip makers (producer–user learning). This began to occur as the subsidiaries required greater support from local material and component suppliers.

Also in contrast with leapfrogging, the TNCs began in pre-electronic activities, mainly assembly and testing. Gradually, firms progressed to more complex engineering tasks, suggesting that the stark contrast between the new (electronics) and old (mechanical) paradigms implied by leapfrogging is misleading. Paradigms appear to overlap with the new building on the old. Furthermore, firms began at the standardized, mature end of the product cycle, rather than the early stage. They then progressed stage by stage to more complex, advanced products.

As firms developed, the government improved the infrastructure, education and supply of engineers and technicians. Incentives were introduced to encourage firms to expand their higher technology activities (e.g. the special tax status for headquarter activities). The EDB responded to human resource needs by improving the engineering and R&D facilities in universities and polytechnics and by setting up special R&D institutes such as the IME. It is unlikely that the TNCs would have upgraded their activities to the extent witnessed without competence and skill on the part of the government.

6.8   THE HARD DISK DRIVE (HDD) INDUSTRY

Sector Overview

During the 1980s Singapore's HDD (or Winchester disk drive) industry grew from humble beginnings to become a multibillion dollar industry. Dominated by US manufacturers who account for around 80 per cent of the world's output, Japanese firms such as JVC, Toshiba, Fujitsu, Matsushita and Sony also compete. Singapore, the world's largest producer of HDDs in 1991 and 1992, focused mainly on 2.5 inch, 3.5 inch and 5.25 inch drives. Also produced were tape drives (0.5 inch and 0.25 inch cartridge) and back-up systems. US producers in Singapore included Seagate, Conner Peripherals,

Maxtor, Micropolis, Western Digital and Archive. Employment in the industry was 25 000 in 1991. Firms began with relatively simple assembly tasks but quickly moved to manufacturing and product development as well as marketing.

Modern HDD production is a large volume, high-precision, automated activity. It requires investments in computerization, continuous engineering and R&D back-up for new product design. Production involves clean rooms and sophisticated testing equipment. Singapore's capabilities in HDDs are reflected in locally designed spindle motors, servo mechanisms and CD-ROM drives. Some firms introduced computer-integrated manufacturing.

History and Choice of Location

Seagate, which began business in the US in 1979, grew to just under US$1 billion by 1987 and US$2.7 billion in 1991. In 1992 it was among the 200 largest corporations in the US, employing 40,000 people in 17 countries. In 1982 the company became the first HDD manufacturer to enter Singapore, making printed circuit boards and other computer-related products. In 1983 it produced the first HDDs locally and by 1991 the Singapore subsidiary employed more than 12,000 people. Conner Peripherals began in the US in 1986 and started operations in Singapore in 1987. By 1991 Conner's three Singapore factories employed more than 4,300 workers with sales of approximately S$2 billion.

The companies chose to locate in Singapore for tax and other incentives, low labour and engineering costs (compared with the US), effective training schemes and efficient air and sea freight systems. Around 60 per cent of total HDD sales were exported to the US, compared with 25 per cent to the EC and 10 per cent to Pacific Asia. Expansion was mainly the result of cost advantages relative to the US. Conner, for instance, found that within three months of plant start-up, Singapore unit costs were already far lower than those of the US parent.

Seagate's Singapore subsidiary grew rapidly through internal expansion and outside acquisition. In 1984 it acquired Grenex (now Seagate Magnetics). Afterwards, it set up a product design and development team for thin-film magnetic media at the Singapore Science Park. Following an expansion of in-house printed circuit board facilities in 1985, it purchased Aeon Corporation (now Seagate Substrates) in 1987 and Integrated Power Systems (now Seagate Microelectronics), a maker of custom power integrated circuits. In 1989 Seagate made two further acquisitions: Imprimis Technology (for storage devices for workstations and mainframes) and Peripheral Components International (for advanced magnetic storage heads). Both Seagate and Conner integrated vertically in order to grow, control input quality and ensure adequate supplies of motors, disks, substrates, heads, media, printed circuit boards and custom semiconductors.

Backward Linkages and Technological Upgrading

Although at the time of start-up there was little local HDD supporting industry, with expansion, foreign materials and component suppliers followed the TNCs into Singapore. By 1991, most of Seagate's materials were purchased locally and some inputs were manufactured domestically. Seagate's early success attracted other US HDD producers to the island, bringing with them more material and component suppliers. Eventually, most US production was located in Singapore, leaving pilot production, design and R&D to the US parent. During the 1980s, supply firms grew up in components and subassemblies (such as precision machined and stamped parts, die-casting and surface-mount printed circuit board assembly), significantly upgrading the local supply chain.

During the 1980s, HDD manufacture became highly automated as a result of the increasing complexity of new products. Computerized control systems and automated surface-mount technology were installed in Singapore to ensure capacity utilization and high yields and to control inventories. Although by 1992 continuous process engineering was widespread, new product design was carried out largely in the US. However, Seagate's local plant worked closely with the US parent on the manufacturability of new product designs. Two or three Singapore engineers were constantly in the US plant.

Conner's main Singapore plant advanced to become its world centre for corporate manufacturing with the local managing director in charge of production operations overseas. Most of Conner's product design was still carried out at the R&D centre in Colorado, though some product engineering had been transferred to Singapore. Conner's production automation systems were designed in-house by staff in Singapore. The Singapore facility was also responsible for transferring new processes to Conner's plants in Scotland, Italy and Malaysia.

Implications for Learning and Leapfrogging

HDD manufacture is a case of rapid technological learning among firms in Singapore. Initially, firms located in Singapore for cost and infrastructural reasons. The first entrants were followed by others, creating a cluster of large companies. Engineers graduated from Singapore and neighbouring countries of the region, forming links with local universities, polytechnics and the government-funded Magnetics Technology Centre. The plants rapidly graduated from production learning to investment-led, innovative learning. This was partly due to the fast-developing nature of the industry. It was also due to Singapore's host country advantages.

In contrast with leapfrogging, the HDD case illustrates entry into a precision engineering industry, partly pre-electronic in character. The vital skills were precision motor and fine machinery engineering, circuit board assembly, electromagnetic computation and electromechanical interfacing. Rather than software engineering (the core skill for microelectronics) the source of Singapore's advantage was its abundant supply of highly skilled engineers.

Firms quickly exploited the infrastructure generated during the 1960s and 1970s, learning to develop new production systems and to transfer these processes abroad. Some grew by acquisition and diversification. Learning-by-imitation occurred once the leaders were established successfully. The HDD case shows how a few leading firms were able to stimulate the growth of a whole new industrial cluster, once the country's infrastructure was adequate.[23]

6.9   CONSUMER ELECTRONICS

Origin and Expansion

Consumer goods firms were the first to become established in Singapore. By 1991 the sector employed around 26,000 people, with sales of S$5.7 billion. The two cases below, Philips Singapore and AT&T, compare the processes of learning within an early and a late entrant. Philips Singapore (a subsidiary of NV Philips of the Netherlands), which began its commercial activities in Singapore in 1951 with a trading office of four staff, started producing radios and PABXs in the late 1960s. In the 1970s it diversified into domestic appliances (mostly flat irons), cassette recorders, TVs and components.

By 1980 Philips produced test automation machines, precision moulds and domestic appliances (especially irons). During the 1980s, the firm added audio equipment, colour TVs, compact disc players and tuners to its product range, while continuing to manufacture precision tools and dies for other Philips factories. In 1991 Philips Singapore employed 6,100 people in five separate factories (*Singapore Electronics Manufacturers Directory* 1992 p. 76). With sales of around S$1.5 billion, Philips was one of the largest TNCs in Singapore.

AT&T Consumer Products Pte Ltd (ACP), a recent entrant, was one of the largest AT&T factories outside the US. In 1991 it employed nearly 4,000 staff and supplied a wide range of products and services. ACP, AT&T's first offshore manufacturing facility, started in Singapore with an investment of around S$100 million in 1986, making corded (standard) residential telephones mostly for the US market.

Technological Advance and Automation

Production lines at Philips became steadily more complex and automated over the decades. By 1992, with operations generally requiring sophisticated engineering support, many graduate engineers and technicians were hired from within Singapore. The tuner factory, the largest Philips's plant of its type worldwide, made four major product families each with a wide variety of designs. Within the computer-integrated manufacturing systems, quality control and testing were computerized.

As in the cases of chips and HDDs, many line workers were still required for tasks not yet automated, such as loading and checking. Philips's video factory was also modern and highly automated. Production was able to switch from one TV model to another in around ten minutes. Local designers used computer-aided design and manufacture to assist in the development of audio products.

As labour costs increased, the companies relocated more mature products to other parts of the region. ACP, for instance, invested about US$5.6 million in a satellite factory in Batam in Indonesia (part of the Growth Triangle) because of labour shortages in Singapore. New products were introduced to the main plant, while labour-intensive operations were relocated to Batam.

By 1990 ACP was AT&T's corporate centre for top-of-the-range cordless telephones, having designed and developed new digital phones in-house. The plant was advanced by world standards, using automated processes and computerization and, as with Philips, requiring high-quality engineering support to ensure the output standards. Processes in use at ACP included surface-mounting (compared with hand-placed) technology, reflow soldering and automatic testing. Lines were able to switch from one type of product to another fairly quickly. The company used computer controlled scheduling, performance measurement, planning, purchasing and shipping.

ACP's local R&D was organized as a division of AT&T Bell Laboratories, providing design support for ACP and OEM suppliers in the region. The local laboratory was responsible for introducing new products, including the cordless telephone noted above.

ACP used OEM suppliers to quickly increase capacity in response to market changes. Under OEM arrangements, strict quality and delivery controls were imposed upon suppliers, including computerized just-in-time systems. Under close business and technological relationships between ACP and its OEM partners, local companies were linked to ACP by computer to relay purchasing and other business information back and forth. By providing engineering support to OEM suppliers, such agreements helped train local firms in advanced manufacturing techniques and modern business practices.

Both ACP and Philips carried out formal training and had formed educational links. As of 1992, as well as running around 70 programmes for basic training, skill upgrading and quality control, ACP contributed towards scholarships at the National University of Singapore and Nanyang Technological University. Both firms were involved in the government's Local Industry Upgrading Programme, working to assist local suppliers improve their production systems, product quality and management efficiency. Philips's first educational links began in 1971 with a four year craft apprenticeship programme organized with the Ministry of Education to supply craftsmen and technicians. In 1975 the EDB–Philips Government Training Centre was established for precision engineering training.

Implications for Learning and Leapfrogging

In contrast with leapfrogging, Philips's skill requirements began with craft and engineering whereas ACP, a late entrant, began at a more advanced engineering level. Both examples point to the importance of precision engineering to production and product design in Singapore's consumer electronics industry.

Philips, over a 30 year period, incrementally accumulated technology, beginning with craft and technical skills and, later, process engineering and product design. As it expanded, Philips's subsidiaries provided more technical back-up, engineering support, design services and computerization of production. Although Philips lagged behind the research frontier set by its parent, it had narrowed the manufacturing and design gap.

The case of ACP (like the HDD examples) shows how a late entrant was quickly able to exploit Singapore's infrastructure, producing high-technology goods, requiring sophisticated design and process capabilities. The ACP plants were among AT&T's most advanced worldwide. While early investors like Philips took three decades to reach this stage, ACP took less than five years, benefiting from the physical and human resource infrastructure built up over the past decades.

6.10   A LATECOMER ENTRANT IN COMPUTERS

Origin and Expansion

In 1992, the Wearnes Hollingsworth (WH) Group was the largest locally owned maker of PCs and peripherals. WH, originally an Australian-owned distributor of British cars, was purchased by the Overseas Chinese Banking Corporation in the 1930s. Beginning by making simple connectors some 30

years ago, WH moved into high-precision stamping, plastic injection moulding and plating in the mid-1970s. Computer sales expanded rapidly during
the 1980s within a subsidiary company, Wearnes Automation, established in
1983. The Group also made Winchester and floppy disk drives, add-on cards
and a range of electromechanical devices such as connectors.

The Group gained know-how by acquiring several small companies in the
US, including a 40 per cent share of Advanced Logic Research and two other
chip design firms. Other purchases included: United Circuits of Hong Kong
in 1989 (a maker of printed circuit boards); and OMEDATA of Indonesia in
1986 (a chip packager for National Semiconductor of the US). Although
some of the acquisitions proved difficult and costly, they helped WH grow
and assimilate US technology. To advise on the move into PCs, a manager
(later a board director) was hired from a relatively small Singaporean venture
previously engaged in printed circuit board manufacture.

The firm's existing electromechanical competences assisted the move into
PCs and peripherals. Building on its core strengths in connector manufacture,
stamping, chip packaging, plastic moulding and electroplating, chip design
capabilities were added through the US acquisitions. Most of WH's R&D
(around 6 per cent of sales) was applied work, oriented towards near market
needs. Engineers up to MSc level were recruited for most divisions, while
MBAs were avoided unless they also held engineering degrees.

Up until the early 1990s the firm operated as a follower in the market,
moving into established areas, acquiring technology as and when necessary.
After 1992 a new division (Wearnes Computer Systems) set up more than 50
service centres in Pacific Asia and Europe to stock Wearnes computers and
provide after-sales support. A total of 200 such centres were planned for
Europe, the US and elsewhere. This event marked a new phase of internationalization and confidence.

Learning Through OEM

Company sales, especially PCs, grew rapidly during the 1980s and WH
earned a reputation as a high-quality, fast-delivery OEM manufacturer of
computer systems, printed circuit boards and components. Senior managers
were hired in from TNCs while most engineers and technicians were recruited from local universities and polytechnics.

Under OEM deals, WH gained from the TNCs' experience in a variety of
technologies, while the TNC buyers distributed and marketed Wearnes's
output under their own brand names. From the point of view of WH, OEM
yielded economies of scale both in production and in the purchase of key
inputs such as microprocessors. OEM also provided a justification for investments in automation technology, improving WH's productivity and quality.

Under OEM the firm learned the rigours of high-quality, fast turn-around production.

Gradually, the firm developed its own in-house designs and brand names, progressing beyond OEM. Nevertheless, as in the Taiwanese and South Korean cases, WH continued with OEM, despite in-house competencies in design and development. For instance, in 1992, the firm agreed to a large OEM arrangement with IBM, the US computer maker. The arrangement offered market prestige and further scale advantages.

Implications for Learning and Leapfrogging

As with the other examples, WH's progress was gradual, rather than a radical jump. Also in contrast with leapfrogging, the company built its strength upon its precision and mechanical engineering skills, rather than information technology skills. WH progressed from low-cost assembly to OEM and then own-brand product design. Capabilities in technology, management and marketing were learned, while US acquisitions provided new skills needed for PC design. On the marketing side, WH began its own brand sales, setting up international sales offices in many countries.

The case of WH confirms that advanced latecomer competencies do not necessarily displace the ones learned earlier. New software design skills augmented older mechanical competencies. Gradually more sophisticated capabilities were accumulated and deployed to maximize the company's competitive advantage.

6.11 SINGAPORE'S TECHNOLOGICAL ADVANCE

Since the 1960s, the TNCs have upgraded their technological capabilities, expanded Singapore's export industry and contributed to full employment. They have trained local technicians, engineers, managers and OEM suppliers. Technology has been transferred gradually, enabling the subsidiaries to narrow the gap between themselves and their parents abroad. As with the other dragons, the direction of learning has been from simple to complex tasks, culminating in the early 1990s with product design and advanced precision engineering for manufacturing. However, industry overall still lacks significant R&D capabilities.[24]

The evidence is sharply inconsistent with the leapfrogging hypothesis. Technology was accumulated in a gradual and painstaking manner, with firms engaging in a hard slog of incremental learning in response to factor price increases and the improving infrastructure. Contrary to leapfrogging, much of their advance was in pre-electronic activities such as mechanical, electromechanical and precision engineering, rather than software or R&D. Also in contrast with leapfrogging, firms tended to enter at the mature, well-established phase of the product life cycle, rather than at the early stage.

As far as industrial progress is concerned, the evidence suggests that old and new paradigms are closely connected and overlap in a variety of ways. To progress towards electronics, companies found it necessary to develop skills in plastics, mouldings, machinery, assembly and electromechanical interfacing, many of which are common to earlier vintages of technology. Furthermore, many of the products exported embodied many pre-electronic inputs. Singapore's route towards electronics was through competence building in basic industrial technologies, rather than by leapfrogging.[25]

For other developing countries, the evidence suggests that, given inducements, TNCs can be persuaded to transfer technology to their subsidiaries. In Singapore the TNCs willingly trained local employees and managers and formed many research and educational connections. This was encouraged by educational policies, the improving infrastructure and the expanding Pacific Asian market.

The pace and direction of technological progress of TNCs in Singapore was similar to that of latecomer firms in the region. Like the latecomers, the foreign subsidiaries learned in response to market opportunities and the abundant supply of low-cost engineers and technicians. Early TNC entrants, as with latecomers, learned by upgrading production processes and eventually by designing new products, graduating from simple to advanced learning over a period of decades. Later entrants quickly advanced to innovative learning. Together with the latecomers, the TNCs in East Asia are a central part of the region's burgeoning economic development.

NOTES

1. For details see Fok (1991 pp. 257–8) for Hong Kong, Sakong (1993 p. 232) for South Korea, O'Connor and Wang (1992 p. 41) for Taiwan, and EDB (1992a p. 10) and *Electronics* (26 April 1993 p. 5) for Singapore (all official figures in current prices). In South Korea electronics accounted for around 28 per cent of total exports in 1991. These figures are not strictly comparable for definitional reasons. The Singapore figures include around 25 per cent re-exports which is more than the other dragons (*Financial Times* 29 March 1993, Singapore Survey p. 11).
2. The research was carried out between 1991 and 1993. Thirty-five interviews were conducted with TNCs, local firms, industry associations, universities and government organizations connected with electronics and R&D.
3. Soete (1985) first put forward the leapfrogging view. Since then it has become common parlance in policy making circles, despite the lack of empirical evidence for the claim. Perez and Soete (1988) argue that windows of opportunity arise at the early stages of the diffusion of a new paradigm. See Perez (1985) for the pioneering institutional work on techno-economic paradigms. The latter study explores the complex relationship between the techno-economic sphere and the socio-institutional framework within which new

paradigms diffuse. Dosi (1982) defines the technology paradigm and provides an introduction to the semiconductor paradigm. Pavitt (1984) was the first to criticize the leapfrogging view, arguing that the diffusion of semiconductors was a complex, gradual
process with the new technology building on the old: a process of creative accumulation.
The terms information technology, semiconductor, microelectronics or electronics paradigm tend to be used interchangeably in the literature. The new paradigm consists of
industries based primarily on semiconductor and software technologies (including the
computer, telecommunication, semiconductor, software and consumer electronic industries). It also includes the application of information technology to conventional industries
such as automobiles and aerospace.

4. See Hobday (1990) for the case of Brazil. Antonelli (1991) provides international evi-
    dence on the relatively fast diffusion of advanced telecommunications in developing
    countries.
5. Archambault (1992) looks at the case of the semiconductor industry in the Republic of
    Korea. He argues that, contrary to the view of Perez and Soete (1988), South Korean firms
    entered into the industry at the mature, not the early stage of the product life cycle.
6. See Sisodia (1992), National Computer Board (1992), Gilbert (1990) and Davies (1988).
    The role of information technology in banking and insurance is discussed by Onn (1991).
7. Singapore had far fewer of the wealthy overseas Chinese entrepreneurs, who led indus-
    trial development in Hong Kong.
8. Interviews with firms and industry associations. For an historical discussion of incentives
    and their effect, see Yue (1985).
9. EDB (1992a) p. 10. Figures are in current prices (the Singapore dollar exchange rate in
    1991 averaged US$1 = S$1.73, and in 1992 US$1 = S$1.63 (*Financial Times* 29 March
    1993, Singapore Survey p. 11).
10. *Computrade International* (1 November 1992 p. 34).
11. *Singapore Investment News* (1991 p. 10).
12. EDB (1992a p. 15).
13. Normally, TNCs would carry out such work in their domestic headquarters. The Institute's
    other main aim was to supply skilled researchers for semiconductor firms (interview with
    Institute for Microelectronics, 1992).
14. NSTB (1991 p. 13). Research scientists and engineers are defined as those with a batchelor's
    degree or above.
15. NSTB (1991 p. 17).
16. Information from the British High Commission, Singapore.
17. For discussion of the problems see Vatikiotis (1993 p. 54).
18. EDB (1992a p. 2).
19. EDB (1992c p. 1). Data for 1992 reported in *Electronics* (26 April 1993 p. 5).
20. Calculated from IEEE (1991 p. 27). The strategies of two of the largest local firms
    (Wearnes and Chartered) are discussed below.
21. TECH and Chartered were recently established, both with close government involvement
    (see below).
22. This partly explains why employment increased as automation proceeded. The expansion
    of plant output also contributed to employment expansion.
23. This form of clustering behind the frontier resembles the Taiwanese experience (Chapter
    5) except that in the Singapore case TNCs, rather than local firms, provided the dyna-
    mism.
24. Overall, Singapore still lagged behind in information-based software skills and R&D.
    Although R&D was the target of government technology policy, it was not the basis of the
    industrial success. Indeed, as Section 6.4 shows, Singapore's capacity in science and
    advanced technology was conspicuously weak compared with OECD countries and other
    East Asian NIEs.
25. As noted earlier, this is not the case for infrastructural leapfrogging, which occurred in
    Singapore and elsewhere.

7. Hong Kong: *laissez-faire* technological development

7.1 MARKET-LED INDUSTRIALIZATION

Hong Kong, a small city state like Singapore, has a population of just under six million, 98 per cent of whom are ethnically Chinese. It is a major financial capital for East Asia and a historical trading route into Mainland China (to which it is due to return in 1997). Of the four dragons, Hong Kong is the only economy to have pursued a *laissez-faire* approach to industrial development.

Although many local firms boast significant design capabilities, Hong Kong lags behind the other three dragons technologically. Nevertheless, it has built up a significant position in the world electronics industry, relying on a mixture of foreign TNCs and small indigenous local firms. By 1990 the city exported in the region of US$7.5 billion worth of electronics products (Fok 1991 p. 257), nearly 60 per cent of the total exported by Taiwan, a much larger economy with a population of around twenty millions.

This chapter explains how the electronics industry developed in Hong Kong, within the *laissez-faire* policies of successive administrations, and how local firms learned their skills and overcame barriers to entry.[1] As in the other cases, the chapter analyses the main international sources and channels of foreign technology, showing how these were exploited by latecomer firms.

Hong Kong's forward march in electronics is closely tied to the opening up of Mainland China. As Hong Kong firms evolved into technologically competent companies they have outsourced much of their manufacturing activities into China, contributing to the burgeoning economic growth of that country.

7.2 THE HISTORICAL CONTEXT

Hong Kong was ceded to the British in 1842 under the Treaty of Nanking (Haggard 1990 p. 116). It served as an entrepôt for British exports and a major outlet for Chinese trade. By the turn of the century some shipping-related industries had emerged, but trade was the main source of income. In 1900 Hong Kong handled around 40 per cent of China's export trade.

162

In 1935 a commission appointed by the Hong Kong Legislative Council supported the ongoing practice of *laissez-faire*, arguing that there was: 'little scope in a Colony like Hong Kong, having no natural raw products and a small domestic consumption, for the ambitious schemes of economic reconstruction or national planning which have become the modern fashion' (Haggard 1990 p. 117).

During the years 1937 to 1941 flight capital from the Mainland arrived in Hong Kong as a result of the outbreak of hostilities. Some of the new capital was invested in the production of light military equipment. However, the Japanese occupation halted business and the city's population fell from around 1.5 million to 500,000.

After the Japanese occupation, the Administration began providing some housing and essential commodities, but otherwise stuck to the *laissez-faire* doctrine. Until 1950 the main source of economic progress was trade with China and the rest of the world. China's civil war led to a rapid influx of immigrants in 1947 and a massive inflow of flight capital. In 1949–50 the annual inflow of capital constituted around 65 per cent of Hong Kong's national income (Haggard 1990 p. 118). With the Korean War, the US closed the border to Communist China in 1950. This led to a further influx of refugees arriving from the Mainland (Vogel 1991 p. 68).

The new arrivals included a number of Shanghai textile factory owners and managers. The Shanghai capitalists imported finance, modern machinery, raw materials and entrepreneurial talent. A sizeable investment took place in the cotton spinning industry in the early 1950s, which, given the tiny size of the territory, was forced to seek overseas market outlets for its produce. Textiles also gave rise to the development of the apparel industry in Hong Kong (Vogel 1991 p. 72).

In the mid 1950s a second wave of light industries was led by the local entrepreneurs. Their growth was helped by the large trading companies who supplied local firms with designs and specifications. Small business owners moved into labour-intensive industries including plastic goods, toys, low-grade electronics, watches and clocks.

Compared with Singapore, Hong Kong entered the 1960s with a much stronger base of local entrepreneurs, managers and capital with which to industrialize. The Shanghai refugees arrived to find Hong Kong banks keen to fund new ventures and to replace the declining entrepôt trade. Local capital enabled some Chinese immigrants with little capital or assets to set up factories. Some set up in Tsuen Wan on the southwest coast of Hong Kong's New Territories to produce goods for export. Many of the refugees were from the Chinese province of Guangdong, one of China's most commercialized regions.[2] With increasing personal income, many Hong Kong entrepreneurs sent their children to study in England, the US and other countries. Many returned with qualifications in business management, finance and engineering. Some learned about the emerging markets for new electronics products (Vogel 1991 p. 72).

7.3 ORIGIN OF THE ELECTRONICS INDUSTRY

The electronics industry began in the late 1950s when a small number of Japanese transistor radio producers entered to exploit Hong Kong's low-cost labour and low taxation rates. Local firms also began importing components to make transistor radios, following a decision by Japan to allow the export of electronic components to Hong Kong (HKPC 1982 p. 8). As a free port, Hong Kong quickly became a spot market for trading in excess stocks of electronic parts (Fok 1991 p. 258). Hong Kong's first electronics company (Champagne Engineering Corporation) began to assemble radios for Sony in 1959. In 1960 it began producing its own radios, undercutting the Japanese (Henderson 1989 p. 80).

American FDI also helped form the Hong Kong electronics industry. As in the other dragons, the low cost of labour attracted US components producers, including Fairchild, Ampex, and Teledyne Semiconductor during the early 1960s. By the mid-1960s almost every large US consumer goods and semi-conductor manufacturer had established operations in Hong Kong. Focusing on the assembly and testing of diodes, transistors, capacitors and other com-ponents, the TNCs provided a training ground for engineers, middle manag-ers and technicians. They quickly learned to benefit not only from cheap labour but also the relatively low-cost, but highly productive, local engineers, technicians and professional staff.

As a result of the these activities, the share of electronics in manufactured exports rose from 2.5 per cent in 1961 to 12.3 per cent in 1976. Thereafter it grew to 19.1 per cent in 1981 to 22.6 per cent in 1987 (Henderson 1989 pp. 81–5; HKPC 1982 p. 9). By 1981 electronics was the fastest growing export sector and the second largest export industry after garments. The latter constituted around 35.2 per cent of exports in 1981.

During the 1970s, US firms stepped up their semiconductor packaging and testing and began producing other computer parts for export. However, the surge of local companies into cassette tape recorders, calculators, electronic watches and clocks enabled them to overtake the TNCs as the main source of production and employment. The total number of electronics firms, mostly local, grew from a tiny number in 1961 to 1,180 in 1987 while employment reached around 79,000 (Henderson 1989 p. 94 and p. 86). The share of FDI in production fell during the 1970s and 1980s as local firms exported large quantities of TV games, PCs, peripherals, digital telephones and many other consumer electronics.

7.4 POLICIES FOR ELECTRONICS

Unlike the other NIEs, Hong Kong maintained its *laissez-faire* approach to electronics and industrial development in general. However, several agencies played an important, sometimes creative role in facilitating electronics progress. The Hong Kong Productivity Council (HKPC), a statutory body, was established in 1967 to provide technological support for local firms. It promoted the application of microprocessors in the late 1970s and, during the 1980s, extended its influence with support schemes for ASICs, printed circuit board technology, mechanical design and reliability testing, the promotion of international standards (e.g. ISO9000) and technological support for SMEs (HKPC interviews 1993).

Although the HKPC is funded mainly by the administration, it also raises finance from consultancy services. With a staff of around 500 in 1993 its budget was in the order of HK$300 million (US$43 million). With professional consultants in engineering, science, business administration and management, most of HKPC's work could be described as business-oriented engineering, training and quality promotion. In 1993 it dealt with around 4,000 firms, operated around 500 training courses and had assisted about 50 local firms to gain ISO9000 accreditation (essential for exporting to Europe and beyond). The HKPC assists SMEs through subsidized training programmes and consultancy, by making available modern electronics facilities and by helping companies to design new products. In 1992, for example, the HKPC organized two consortia, one of 15 firms to design a palmtop computer, another of 8 to produce a cordless telephone. Each member firm contributed some element of the design to the new models. According to the HKPC these projects were a success, bringing new innovations to the market and helping small firms overcome their size constraints (HKPC interview, 1993).

While it is not possible to assess the overall impact of the HKPC, it assisted some local firms to overcome entry barriers and realize new market opportunities. It also coordinated policy discussions on the future of the industry, bringing together industry leaders, academics and administration officials.

Other significant support bodies include the Vocational Training Council (VTC), the Trade Development Council (TDC) and the Industry Department (ID). The VTC is a statutory organization which operates eight technical institutes and two industrial training centres, providing craft and engineering courses in electronics and related industries. The VTC provides short training courses in machine shop and metal working, plastics, precision tooling, CAD/CAM and other subjects, whereas the TDC is responsible for promoting overseas trade. The TDC brings together industrialists, trade associations, senior administration officials and organizes trade fairs. It also operates a computerized trade enquiry service, providing product information by company for local and overseas buyers. The ID attempts to promote industrial investment by providing prospective investors with information on suitable industrial locations, labour searches, and other investment factors.

As with the other dragons, Hong Kong's educational institutes supply graduate and undergraduate education in electronics. Some, such as the University of Hong Kong and the new University of Science and Technology are heavily engaged in promoting R&D for electronics and advanced information technology. The Chinese University of Hong Kong spun off various companies, including a leading manufacturer of specialized liquid crystal displays.[3] Also two local polytechnics carry courses for electronics and related disciplines, some of which are operated by the VTC.

7.5   PERFORMANCE IN ELECTRONICS

Table 7.1 shows product exports for 1984 and 1990. Electronics exports doubled from around HK$16 billion in 1982 to HK$30 billion in 1984, thereafter growing steadily to nearly HK$60 billion in 1990. Over the period 1984 to 1990, the share of watches and clocks in total exports remained stable at just over 20 per cent, reflecting Hong Kong's agglomeration of skills and marketing expertise in the consumer electronics field. Since 1984 the

Table 7. 1   Hong Kong exports of electronic goods (excluding pens/watches) (HK$ billions)

	1984	Share (%)	1990	Share (%)
Watches and clocks	6.2	20.7	13.1	22.4
Components	4.5	15.0	16.3	27.8
Radios	4.8	16.0	1.2	2.1
Computer parts	6.1	20.3	9.6	16.4
Telephones	1.6	5.3	1.2	2.1
Parts for radios	1.5	5.0	3.1	5.5
Video tapes	0.3	1.0	1.5	2.5
Computer peripherals	1.5	5.0	3.7	6.2
TVs	0.4	1.3	1.6	2.7
Other consumer	3.1	10.4	7.2	12.3
Totals	30.0	100	58.5	100

Source: Calculated from HKPC data, cited in Fok (1991 p. 261).

share of radios in total exports fell by 14 per cent as a result of the maturation of the market and the transfer of low-end production to China. Conversely, component production increased by more than 12 per cent, becoming the largest sub-sector. This was partly due to the expansion of semiconductor assembly by Motorola (see below) and other firms.

In 1990 the main product lines (in order of magnitude) were components, watches and clocks, computer parts, miscellaneous consumer goods and computer peripherals. In addition, there was production of PCs, electronic games, printed circuit boards, TV receivers and liquid crystal displays.

Most local firms produced under OEM arrangements for large Japanese and US firms, although some had developed their own brand names. OEM was the dominant activity in watches, clocks, printed circuit board assemblies and telecommunications peripherals. Exceptions to the OEM rule included Porro Technologies, which exported its own-brand workstations to Australia and Pacific Asia. Other firms had notable design successes in fax machines, liquid crystal displays, cordless telephones and small-screen colour TVs (interviews, 1993). In low-end products several firms have developed their own international market outlets, selling under own-brand names.

As in the other NIEs, Hong Kong is no longer a rudimentary OEM exporter. A large proportion of consumer goods exports embody significant local design content. Some detailed examples are provided in the case examples below. Most Hong Kong firms have taken control of production processes and many have made significant strides into design, both independently and in collaboration with foreign ODM partners.

7.6 INDUSTRY STRUCTURE AND ADVANTAGES

Like Taiwan, Hong Kong's industrial structure in electronics is dualistic, comprising many small companies and a few large TNCs. Eighty per cent of local firms employed less than 50 staff in 1990 and only one employed more than 2,000 people (in Hong Kong).[4] Although the largest local companies are substantially smaller than Taiwan's leading firms, many grew rapidly during the 1980s. The five largest Hong Kong latecomers contributed around US$1.1 billion to output in 1989 compared with US$2.6 billion in Taiwan.[5] A selection of major electronics exporters is presented in Table 7.2.

Although smaller than local industry, foreign firms remained an important contributor to electronics exports through the 1980s and into the 1990s. In 1989 total FDI in electronics amounted to HK$8.6 billion (just over US$1 billion) around the same level as in Singapore (Fok, 1991 p. 271). The US was the largest investor accounting for 55 per cent of the total, followed by Japan with 23.7 per cent, the Netherlands with 9.1 per cent and others 12.2

*Table 7.2   A selection of major electronics exporters in Hong Kong 1992,
sales (US$ millions)*

Video Technology	US$561
Kong Wah Group	US$357[a]
Semi Tech Microelectronics	US$250
Tomei Industrial	US$250
Wong's Industrial Holding	US$220
Conic Investments	US$200

Note: [a] Discussed in Section 7.8.

Source: Annual reports, company interviews and author's estimates.

per cent. In 1992 there were 112 foreign-owned electronics companies, employing around 32,000 people, just over one third of the total workforce. Many of the TNC subsidiaries had taken on some of the characteristics of local firms, functioning as if they were OEM/ODM suppliers to the parent abroad. Some of the TNCs had gained substantial design autonomy in fax machines, PCs, printers and so on. Others competed with rival OEM subsidiaries of parent companies located elsewhere in East Asia.

The growth of Hong Kong's leading firms during the 1980s mirrors the latecomer advance in Taiwan and South Korea. The largest local producer, Video Technology (or VTech), began life as a TV games manufacturer in 1977. It was founded by two friends, both former engineers at NCR in the US. Of VTech's nine board directors in 1993, most had technical experience in TNCs including NCR, Philips, Fairchild Semiconductor and Toshiba. Group sales grew rapidly through the early 1980s, reaching US$164 million in 1988, thereafter jumping to US$263 million in 1991 and US$561 in 1992 (annual report and interview, 1993). Like many other local firms, VTech progressed from consumer electronics into professional systems. By 1992 sales of PCs reached US$362 million, 65 per cent of total sales, compared with US$127 million (23 per cent) for electronic toys and games. In 1993 the firm employed around 10,000 employees, mostly working in factories located in China.

Like many other Hong Kong latecomers, VTech began with OEM for simple goods. By 1993 it had established substantial own-brand and ODM sales. Most product development and design work was located in Hong Kong and the US, its largest export market. Buyers were a major source of technical information, particularly during collaborations in design projects during the 1980s. Suppliers of chips (such as Intel and TI) also helped VTech to learn technology. In 1993 VTech carried out its own ASIC design work in

Hong Kong and had begun selling computers into China. However, Asia Pacific only represented around 6 per cent of sales, while North America accounted for 80 per cent, and Europe around 14 per cent.

Over the 15 years of its operation, VTech like many other companies shifted from simple to complex technologies, reduced the proportion of OEM and achieved some success in establishing its own brand exports. VTech also had the capability to design its own PCs.

One of VTech's main advantages was its relatively low engineering costs, compared with the US. The other was the extremely low labour costs available in China. While comprehensive comparative wage costs were not available, a benchmark study conducted by one major TNC estimated that the entry level cost of a qualified engineer in semiconductors in Hong Kong was around US$25,000 to US$30,000 in 1992, compared with around US$44,000 to US$45,000 in the US. Also technician cost were much lower, with entry salaries of around US$13,500 in Hong Kong compared with US$25,000 in the US. Although engineering and technician costs were high by historical standards due to scarcity and full employment, the Territory still retained a cost advantage over the US and Europe. According to the benchmark study, local engineers were also more productive than their counterparts in the advanced countries.

On the negative side, unlike VTech, most firms were too small to market their own products or to conduct R&D or substantial product design. At an estimated 0.5 per cent of GNP, Hong Kong's R&D spend was the lowest out of the four dragons in 1991 (*Business Week* 30 November 1992 p. 67). Also, unlike the other three, electronics lagged behind the garments sector and the Territory had yet to make much inroad into advanced semiconductors or the more complex electronics systems.

As in the other dragons, backward linkages from electronics stimulated a variety of supply industries, including plastic casings, metal parts and plating, tools, printed circuit board assembly, metal working, materials and components. In total the backward linkage industries employed in the region of 30,000 workers, more than one third of the employment in electronics.[6] Typically the linkage firms were very small with an average of ten or so employees. Nevertheless, they formed the unseen infrastructure of industrial progress. For example, there were around 1,000 mould producing and die-making job shops employing around 10,000 workers in 1992. In metal plating, some 800 firms employed around 15,000 people, while roughly 300 die-casting job shops employed a further 5,000 workers.

Hong Kong's electronics firms benefited from their speed and flexibility. Firms moved rapidly in response to new market opportunities abroad, rather like the latecomers in Taiwan, but in contrast with the large firms in South Korea. Like Taiwanese firms, the owner-founder direction of first generation entrepreneurs allowed decisions to be taken decisively and quickly. Local firms were highly cost conscious and continuously striving to achieve productivity gains, new sources of low-cost components and additional cost advantages from investing in China.

7.7   EXPANSION INTO CHINA

As wages rose in Hong Kong, large numbers of local firms relocated assembly and low-end operations into China, as did their Taiwanese counterparts.[7] By 1992, around three million Chinese in the Guangdong Province worked in Hong Kong-run factories (*Electronics* April 1992 p. 20). This compares with a manufacturing workforce of less than 700,000 in Hong Kong in 1992.[8] Hong Kong firms began to invest tentatively with the opening up of China in the early 1980s, under the new modernization policy. Initially, the aim was to exploit the vast reserves of low-cost labour. Later, firms benefited from low-cost technicians and engineers.[9]

Since Deng Xiao-Ping initiated the reforms in 1978, China's real GNP grew by almost 9 per cent per annum, doubling the size of the economy every eight years (Thoburn et al. 1991 p. 44). However, one third of China's state-owned enterprises showed no return on investment. By contrast, private investments from Hong Kong and Taiwan flourished. Joint ventures grew from less than 100 in 1980 to nearly 40,000 in 1992 (*Telecom Sources* 1993 p. 77). The Guangdong Province economy expanded at around 30 per cent per annum in the early 1990s, fuelled by investments from Hong Kong, its nearest neighbour. Within the so-called fifth Asian dragon, the main industrial centres were Shenzhen and Dongguan.

At the start, the quality of electronics output was generally poor and concentrated in low-end, simple manufactures. According to Fok (1991 pp. 262–3) discipline and lack of motivation were commonly heard complaints among Hong Kong investors. However, during the latter part of the 1980s, initial teething problems were overcome and most firms found workable solutions to their difficulties. Many more manufacturers flooded into China and by 1992 it was estimated that around 90 per cent of Hong Kong firms owned or operated factories or sub-contracted work to plants in China.

While Hong Kong was China's largest foreign investor, according to Thoburn et al. (1991 p. 50) FDI during the 1980s was only marginal to China's overall growth. This is to be expected and follows the pattern of Taiwan and South Korea (see Chapter 2). However, the low aggregate figures mask the central role of FDI in transferring technology, leading to joint ventures, facilitating corporate learning, originating new sub-contract and OEM arrangements and providing access to overseas markets.

Most of the electronics companies in Guangdong were managed from Hong Kong and linked into the world markets through the Territory's extensive trading and marketing networks. Increasingly, Hong Kong firms utilized not only China's cheap labour but also the low-cost skilled software engineers and technicians.

The accelerating process of overseas Chinese investments in the Mainland represented an industrial return home for many former refugees from Guangdong, and their descendants. Their investments in China made a major contribution to the economic growth witnessed during the 1980s and early 1990s.

In spite of political uncertainty over the Territory's return in 1997, most businesses were confident of improving profit opportunities as a result of the synergy with China and the growth of the Mainland market. One survey of Hong Kong's leading local firms and TNCs conducted in 1993 showed that most were very enthusiastic about business relations with China, despite the political difficulties and the overheating of the Mainland economy. The survey covered 176 firms engaged in more than 800 projects valued at around HK$512 billion (around US$73 billion), itself a telling statistic. The report confirmed the considerable scale of Hong Kong–Chinese industrial integration (*Sunday Morning Post*, China Business Review, Hong Kong 30 May 1993 p. 1).

7.8   STRATEGIES FOR INNOVATION: KONG WAH

Company Overview

Kong Wah, the holding company of a group of electronics manufacturers, grew from a tiny operation to a medium-sized TNC. In 1992 the group employed 5,150 people and exported around HK$2.5 billion (or US$357 million), of which around 50 per cent was exported to Europe, the balance to China and Malaysia. TV sales accounted for most of the company's revenues (around 88 per cent), with audio and telecommunications making up the balance. In 1992 the company was the largest TV producer in Hong Kong. In 1993 it began its first overseas production of flat screen TVs and digital stereo TVs in Tyneside in the UK to supply the European market.

In 1993, the Group operated three factories in China and one in Malaysia. Higher quality goods were produced in Malaysia due to the more advanced skills of the local workforce and the relative ease of sourcing components, compared with China. China, by contrast, offered very low labour, engineering and land costs. Engineers were nine or ten times cheaper in China than in Hong Kong. The company found that many low-cost components for mature products could be sourced in China, including picture tubes, filters and coils.

Chinese and Malaysian operations also secured for Kong Wah most favoured nation status, promoting exports to the US and helping to circumvent anti-dumping restrictions from the (then) European Community (EC). Table 7.3 illustrates the geographical spread of Kong Wah's activities.

Table 7.3   Kong Wah: distribution of employment 1993

Hong Kong	600
Schenzen	2,500
Other China	1,000
Malaysia	900
Other[a]	150

Note: [a] Including EC.

Source: Interviews, 1993.

Origin and History

The group was founded in 1979 by a number of local investors, initially for trading in audio products. However, in the first year of incorporation it formed a joint venture (Shenzhen Konka) in China to manufacture simple audio goods. The first orders were sub-contracted from Conic, a major Hong Kong company, which provided assistance with plant start-up and technical support. Only in 1984 did Kung Wah set up a marketing department to promote own-brand sales.

Table 7.4   Kong Wah: product and technology milestones

1979	established as trader/sub-contract manufacturer
1984	began to manufacture own audio products
1985	developed own 4.5 inch black and white TV
1985	developed its first 13 and 14 inch colour TV
1986	began development work on a tranceiver[a] for telecommunications
1986	introduced 19 and 20 inch colour TV models
1988	introduced first tranceiver
1988	developed 26 inch flat-screen TV
1991	developed a 28 inch colour TV

Note: [a] A tranceiver is a device for transmitting and receiving radio signals.

Source: Interviews, annual report.

Mechanisms of Learning

Through the 1980s, substantial technology transfer occurred under OEM arrangements as Kong Wah worked with customers to meet their design specifications. Know-how was also gained from input suppliers, particularly the large semiconductor suppliers from Japan and Europe. In the late 1980s OEM gradually transmuted to ODM as the company offered to produce designs for its customers and began work on its own-brand goods. In the early 1990s, Kong Wah learned new skills in telecommunications when it worked with Amstrad of the UK to develop low-cost satellite receiver products for the mass market. By this time, the supply of detailed design specifications was expected by most large buyers of consumer goods, as in the case of Taiwanese computer manufacturers (Chapter 5).

Kung Wah learned gradually, first by manufacturing simple products, then by designing goods, often in conjunction with buyers. Once a foundation of production and design skills was assimilated, greater efforts were then made to search for new markets and to acquire more complex and diverse technological capabilities.

In 1992 around 91 per cent of the company's business was conducted under OEM/ODM, while 9 per cent was sold under the group's own-brand names Onwa and Konka. By that time most output for overseas buyers was partly or wholly designed and developed in-house by Kong Wah.

Coupled with internal engineering investments, the OEM/ODM system provided a variety of opportunities for acquiring technology. In some cases, large US or Japanese trading companies would approach the group to purchase a specific product, usually with the aim of securing low-cost production. The client's engineers would then work with Kong Wah's on the product's initial conception, then the prototype and sometimes the capital equipment and production facilities. Some hands-off OEM/ODM clients supplied the overall design specification, leaving Kong Wah to organize the design of the printed circuit board layout, the necessary tooling, the facilities for testing and the planning of all necessary mechanical and construction work. Typically, the importer would employ its own engineers to advise on large projects and later to provide the after sales servicing and maintenance in the client's market.

In other cases the Group's own engineers would initiate an idea for a new product either to be marketed by Kong Wah, or to be sold to an ODM customer. To overcome the lack of complex software skills in Hong Kong, in 1989 the company set up a joint software development venture in Europe. The partnership covered specialized areas of software engineering (e.g. for tuning systems) difficult to find in Hong Kong, but necessary for developing new features and functions.

Technological Resources

In 1992 105 R&D staff were employed, of which around 60 were engineers, the balance technicians, draughtsmen and administrative staff. The R&D department was located in Kong Wah Video Engineering Ltd, one of the Group's companies. Rather than blue sky research, most of the R&D staff were engaged in designing new products, improving product quality and features, and purchasing and modifying capital equipment. Most R&D work was oriented towards the short to medium-term needs of customers, carried out in Hong Kong, close to the marketing and administration headquarters. Substantial process engineering work was also carried out within the individual plants.

Group spending overall on R&D, production engineering, design, and the development of moulds and tooling equipment in 1991 amounted to around HK$33 million (US$4.7 million) or 1.3 per cent of sales. While this amount is low by the standards of electronics leaders and followers, as a latecomer, the strategy of Kong Wah was to share development costs with clients and users, and to tightly focus technical efforts on marketable products. Selective, focused engineering was sufficient to build up capabilities incrementally.

Diversification of Market and Technology Channels

Prior to 1988 Kong Wah sold mainly to a small number of large American OEM clients. After this period it established new outlets in Europe, China, Western Europe, Poland, Yugoslavia, Turkey and Indonesia. Frequently, major buyers maintained offices or representatives in Hong Kong, reducing the company's need to invest in distribution arrangements overseas. In 1993 the company employed around 50 people in sales and marketing in Hong Kong. The OEM/ODM clients included many well-known brand names such as RCA, General Electric, Bush, Emerson, Daewoo and Akai. After 1988 Europe overtook the US as Kong Wah's largest export market. In 1991 Germany was the single largest export market, accounting for HK$570 million compared to the US with HK$205 million.

Own-brand sales proved difficult, as with latecomers in other countries. OBM often required the firm to take responsibility for after-sales service, which in turn necessitated setting up offices in the final market. ODM was an alternative which allowed Kong Wah to maintain a convenient division of tasks. Clients controlled distribution and after-sales service, while Kong Wah focused on manufacturing and design.

From Latecomer to Fast Follower

By 1993 the company pursued a fast-follower strategy in some electronics products. It had plans to develop satellite antennae, 900 megahertz cordless telephones (then leading edge), high-definition TV for European Market (possibly under licence) and other systems. While behind the innovation frontier set by the Japanese market leaders, Kong Wah had all but caught up and had become adept at initiating low-cost production in China and Malaysia. However, in marketing functions, Kong Wah remained a latecomer according to the classification in Chapter 3.

Like many other latecomers, as Kong Wah approached the frontier it introduced new mechanisms to continue expansion and acquire more complex technologies. Direct investment into the UK was used to overcome EC trade restrictions. A joint venture in software provided access to complex new skills. In-house investments in design and development further expanded the company's capabilities.

7.9 RJP: CLOSING THE TECHNOLOGY GAP

Origin and Achievements

RJP, a consumer electronics manufacturer, was founded in May 1971 with nine employees. By 1993 annual sales were in the order of HK$400 million (US$57 million). The firm employed around 3,200 staff (mostly located in China) and had sales and service centres in Los Angeles, Chicago and Vancouver (40 per cent of sales were to the US). Its product range included pocket and desk-top electronic diaries, pagers, translators, electronic games, keyboards, karaokes, calculators, simple medical electronics, car radio cassette players and watches. By 1993 the firm had acquired electronic systems and chip design capabilities. It had narrowed the technology gap in consumer electronics and had launched several own-brand products.

Of the company's three founding brothers, one was a trader and one an engineer. The three began the company by investing family savings and some borrowed money from backers in Hong Kong. In 1973 RJP began to assemble calculators, an important activity until 1979. Table 7.5 shows the evolution of key product lines and technological milestones. In 1975, for example, the company began using microprocessors in small volumes to develop simple electronic toys.

Table 7.5   RJP: key products and technological milestones[a]

1971	company start up, production of light dimmer switches
1973	entered calculator production
1975	began making digital watches and TV games
1975	*began use of microprocessors (MPUs)*
1977	*started joint chip design work with TI for radio controlled toys*
1979	entered educational toy market
1981–86	*designed electronic rulers, using MPUs*
1986	began production of medical devices
1986	*designed pocket computer for manufacture (OEM and ODM)*
1991	expanded audio range, added karaoke systems, electronic diaries and directories
1993	*designed palmtop computer*
1993	product range reached 200 entries

Note: [a] Milestones are in italics

Source: Company interviews.

Learning under OEM and ODM

Like many East Asian firms, RJP began manufacturing simple goods under OEM and gradually moved on to increasingly complex electronics. By the early 1990s it had diversified into the lower end of the computer industry offering own-designed databanks and pocket computers. Under OEM, the firm learned how to manufacture a variety of goods. J.C. Penney, the US retail outlet, assisted with basic manufacturing know-how in the early stages by sending quality control engineers to help ensure quality, delivery and productivity. Other local and foreign traders based in Hong Kong provided export outlets and market and technical information.

Through time, RJP increased its contribution to the overall design content of most of its product lines. It experimented with new product models and worked closely with several US chip suppliers. TI's sales engineers, an important source of chip design technology, expanded component sales and purchased jointly developed finished calculators under OEM/ODM. With increasing design demands, RJP hired in some of TI's engineers. RJP's relationship with TI endured and deepened as both companies gained from the relationship.

By the mid-1980s the company controlled the design of most of its key product lines and had registered several patents in the US. As a medium-sized company, most investments were for near market engineering needs, rather than R&D. In 1993 a little research had begun, mostly for future designs for databanks, musical keyboards and educational goods. The number of engineers in Hong Kong rose from 20 in 1986 to 40 in 1993 as a result of general business expansion and increasing technical demands.

RJP's Move to ODM and OBM

During the 1980s RJP progressed from OEM to ODM to OBM, selling more goods under its own brand name and capturing a larger share of the post-production added value. From 1981 to 1986 RJP designed a combined ruler/calculator/digital watch which proved to be a cash cow. After the demand for rulers slowed in 1986, the firm diversified into pocket databanks. This remained a large revenue earner, along with electronic games and toys. In 1986 RJP designed and launched its own-brand electronic diaries and pocket computers in response to expected market demand. This was followed in 1993 by an OBM palmtop computer as RJP attempted to diversify into the low end of the professional electronics market.

OEM buyers frequently imposed restrictions on RJP. Some agreements were exclusively with one purchaser, prohibiting sales into other markets. On the positive side, some OEM/ODM clients (often small US firms) came with new product ideas to RJP to jointly develop experimental systems. This was an important means by which RJP was able to focus its learning on export market needs. OEM and then ODM forced RJP to continuously improve product quality as single shipments of sub-standard goods or delays could lead to the permanent loss of a buyer to another Hong Kong competitor.

Transfer of OEM to China

RJP's two main offices in Hong Kong dealt with finance, marketing and sales. All manufacturing operations had been transferred to China to take advantage of low-cost labour and land. Twelve separate manufacturing facilities were operated in Schenzen and Guangdong. Like many other small Hong Kong producers RJP had learned to couple overseas market demand with the supply of low-cost production capacity in China. In order to benefit from the opening up of China, RJP set up joint ventures with several Mainland enterprises including Softron, Zhenbao, Starmate Garden and Hunan Wire. Softron was a software engineering firm established by professors from the faculty of Guangdong University. Their low-cost software engineers enabled RJP to carry out more engineering work than would otherwise have been possible in Hong Kong. Zhenbao operated a large retail network throughout China, promising future growth for RJP.

RJP: Interpretation

As with Kong Wah, RJP's learning evolved with market expansion. RJP closed much of the technology gap with leaders, progressing from mature to earlier stages of the product life cycle, from simple manufacture to complex design work, and from assembly tasks to process engineering and development. During the 1980s RJP internalized systems abilities and chip design capabilities, enabling it to offer new product innovations under ODM and OBM.

Under the OEM/ODM system RJP's engineers worked hand-in-hand with foreign buyers and components suppliers. Through time the company expanded its competencies and introduced minor product designs and incremental changes to processes. Like other latecomers, RJP embarked on purposeful innovative capacity building in order to pursue profit opportunities in export markets. RJP remained dependent on OEM for some of its sales and to that extent was still a latecomer. However, with OBM the firm had begun the transition from latecomer to follower, competing with improved designs and new proprietary features.

7.10   MOTOROLA: TNC TECHNOLOGICAL INTEGRATION

Start-up and History

Motorola Semiconductors Hong Kong Ltd was founded in 1967 with an office of just two staff. It is a wholly owned subsidiary of Motorola Inc., which had a corporate turnover of US$13.3 billion in 1992. The Hong Kong operation is the headquarters of the corporation's Asia-Pacific Semiconductor Products Group.

Motorola is an example of a TNC which took root in Hong Kong forming many technological linkages within the territory and the wider region. It aims to substantially revitalize its operations through its East Asian activities. According to the company's official mission statement the Motorola Semiconductor Asia Pacific Group: 'maintains a fundamental belief – that the meticulous, productive and strategic mentality of the East, combined with innovative, assertive and flexible thinking of the West, together create a synergy of ideas and working practices which in turn promote advanced technology and provide total customer satisfaction worldwide' (*Motorola Briefing* January 1993). The company competes directly with Japanese and other US companies in Pacific Asia, some of which are also following a path of East Asian revitalization and learning.

The East Asian vision was formulated by a Hong Kong-born senior vice president and general manager (C.D. Tam, MBE). In 1986 Mr Tam became the first ethnic Chinese staff member to achieve such high executive responsibility in Motorola. This high-level fusion of US corporate strength with overseas Chinese entrepreneurial expertise is a formidable long-term challenge to Japanese firms in the region. Similar alliances have taken place in other non-Japanese TNCs (e.g. Philips and TI in Taiwan) and in many looser US alliances under ODM and other partnerships.

Within Motorola Inc. around 52 per cent of sales were to North America in 1991, compared with sales to Europe of 21 per cent, Japan 9 per cent, (other) Asia-Pacific 9 per cent and the rest of the world 9 per cent. Sales to non-Japan Asia reached US$1 billion in 1992, a small proportion of Motorola Inc.'s overall activity, but as large as Japan and nearly as large as Europe, and faster growing than both (interviews, 1993).

Through the 1970s the chip assembly and testing activities were expanded and upgraded from older to less mature products. In 1982 a large production and test facility was set up in Hwai Fong. This was followed by chip design work in 1986 and the automation of semiconductor assembly in 1988. In 1990, the so-called Silicon Harbour Centre was launched, as part of the East Asian resurgence of Motorola. This investment, which cost around US$400 million, began a new complex of chip assembly facilities, design activities, marketing and headquarters. Design facilities were upgraded to include high density memories, ASICs, bipolar-MOS circuits, and other complex chips. Advanced CAD/CAM facilities were introduced to meet the needs of fast moving, short life-cycle chips widely needed among local firms and large TNC buyers.

Technological Status *circa* **1993**

Employment in the Asia-Pacific Group was 2,148, of which 1,198 were engineers, professionals and support staff in 1993. The engineering contingent included around 200 product engineers and a further 200 technicians. Other employees were mainly direct labour, employed in the firm's two Hong Kong manufacturing plants. The Group operated 13 factories in nine countries including China, Singapore, Taiwan, South Korea and India. It was the largest supplier of digital cordless telephones in the region, and ahead of Japanese competitors (e.g. NEC) in China in the cellular communications market.

By 1993 Motorola carried out product marketing, sales, equipment manufacture and associated financial and management operations. However, the main task was semiconductor manufacture. Around 300 engineers were employed in the chip division, of which 210 were production engineers and 90 were chip designers. Design and development activities supported customers within the region with mainstream technology (e.g. LSI, MOS), but not leading-edge circuits. Chips designed locally included gate arrays and full-custom chips for the consumer goods market. No white collar R&D was carried out, but some applied systems development work was in progress (e.g. for liquid crystal display drivers and action matrix TV displays). Systems development was usually linked to the US headquarters where most of the R&D was still carried, as in the case of semiconductor firms in Singapore.

The chip design team had increased in size from around 20 engineers in 1987 to 90 in 1993; a sizeable team, but not in the league of US or European firms. A further 30 or so design engineers were located in Taiwan. The company had little difficulty recruiting local engineers, although four or five of the Hong Kong staff had been educated abroad.

Taking Root

Motorola had formed many backward technological linkages with large clients in East Asia (e.g. Apple, Samsung and Goldstar) to help compete with Japanese, US and some European TNCs. For example, the firm worked jointly with Apple to design 13 inch computer monitors. Larger users developed their key components with companies such as Motorola, requiring technical support to translate a system's design into a silicon chip. Regional growth had accelerated user–producer collaboration in electronics.

The growth in backward linkages (or taking root) on the part of Motorola was not atypical of leading TNCs.[10] Motorola had developed connections with both TNCs and latecomer manufacturers, including several large Hong Kong consumer goods makers. VTech, discussed earlier, was a major user of Motorola's full custom chips. Technology cooperation occurred in electronic games, pocket dictionaries and so on. Motorola also worked with two large monitor producers in Taiwan. Among Motorola's local designs was the DragonKat, a microprocessor which received the Hong Kong Governor's award for industry in machinery/equipment design in 1989. Another was the PocSec chipset, an innovative microprocessor used for pen-based palm top computers with communications facilities.[11]

Motorola's technology upgrading was encouraged by low-cost production engineering, software and design talent in Hong Kong and other parts of the region. In Malaysia, the company's once labour-intensive facility had become sophisticated and automated. Engineers at the Motorola Penang plant designed the Handi-Talkie, a miniature two-way radio which eventually sold 1.5 million units.

Taking root also occurred in less direct ways. Motorola had formed long-term collaborations with local universities in order to recruit graduates and to promote its corporate citizen status in Hong Kong. Many of the company's managers were from local universities and polytechnics. The general manager in 1993 was chairman of the Electronics Committee of the Industry and Technology Development Council of Hong Kong and several other organizations, including the Industry Department, the Technology Review Board and the Hong Kong Productivity Council.

In 1993 the company planned to further integrate its technological activities, primarily to capitalize on the growth of the Pacific Asian market, especially China. In 1993 it began building a US$100 million plant in Tianjin (the first American chip plant in China), which was to make pagers and mobile telephones for the local market, as well as semiconductors. The company also intended to further develop the Silicon Harbour Centre into a major regional facility, while a new chip design innovation centre in Singapore was due to start in 1994.

Interpretation

Like other leading TNCs in the region, Motorola's learning progressed from labour-intensive assembly and testing to design and systems development work. The subsidiary narrowed the gap with the parent company, but like the TNCs in Singapore, had not yet begun substantial R&D or radical design innovation. Despite the grand plan for the region, the Asia-Pacific Division as yet carried out no wafer fabrication, unlike the TNCs in Taiwan and Singapore, and the *chaebol* of South Korea. This, in part, reflected the strategy of the company, and the lagging technological status of Hong Kong, compared with the other three NIEs. Nevertheless, the company had taken root in a variety of ways. Learning was proceeding apace and the diffusion of technology from Motorola to latecomer firms was evident.

7.11   VARITRONIX: A UNIVERSITY SPIN OFF

Origin and Achievements

Varitronix, a small high-technology start-up, was formed in 1978 by eight scientists and engineers, six of whom were from the Chinese University of Hong Kong. Varitronix supplies customized liquid crystal displays (LCDs) and LCD-based electronic systems. LCDs are used in large-volume consumer electronics and low-volume high-specification professional goods. In Hong Kong LCD demand started up in the early 1970s and took off during the 1980s.[12]

Since incorporation, the firm grew by roughly 20 to 30 per cent per annum. By 1993 employment had risen to around 800 with sales of more than US$45 million and client numbers in excess of 1,000. The company operated three manufacturing facilities, located in Hong Kong, China and Malaysia. The Hong Kong and Chinese plants produced sub-systems and dot matrix modules, while the Malaysian factory carried out fully automated volume production using state-of-the-art clean room facilities. Overseas sales offices were set up in 1983 in Los Angeles.

Technological Learning and OEM/ODM

Table 7.6 lists a selection of the company's key developments since start-up. Varitronix began by supplying the surge in demand for LCDs for digital watches in the late 1970s. Among the directors were engineers with doctorates in LCD technology from US universities. At that time, only Fairchild was making LCDs in Hong Kong.

OEM, an important method of learning, began with simple assembly for LCD components, modules and peripherals. In the late 1980s the company was increasingly expected to design and construct modules for clients,

Table 7.6   Varitronix: company milestones and key technological events[a]

1978	establishment of company in Kwun Tong, Hong Kong
1979	*first production of LCDs*
1980	new premises purchased for production in Hong Kong
1981	acquired rival LCD operation
1981	*diversified from mass market to specialist LCDs (e.g. large area displays)*
1982	*added deep temperature displays for industrial applications*
1983	began production in Guangzhou, China
1983	opened sales office in Los Angeles
1983–7	*modified capital equipment*
1985	transferred Chinese operation to Sha Wan, close to Shenzhen
1990	relocated most labour intensive production to China
1990	expanded high end capacity in Hong Kong
1991–2	*patented new design of touch sensitive, portable hand held terminal*
1990–91	opened sales offices in the UK and France
1992	*joint venture with GEC (UK) to develop supermarket checkout displays*

Note: [a] Selection of key technological events in italics (as defined by company directors).

Source: Company interviews and annual reports.

mirroring the shift to ODM shown in other cases. OEM was a means of learning how to integrate LCDs into complete systems, skills which the company initially lacked. Some customers jointly financed the development work for specific devices, enabling Varitronix to use the client to acquire know-how. As many orders were customized, clients often knew their precise requirements. Both OEM and ODM operations were valued as mechanisms of technological learning.

Another channel of learning was in-house changes to imported equipment. Varitronix found that it was sometimes difficult to acquire ready made machinery and tooling. Company engineers worked to modify imported equipment and, in some cases, designed their own systems. Several furnaces for chip manufacture were purchased and modified to suit LCD needs, as no ready furnace on the market was suitable at the time. Similarly, it modified machinery used in the printed circuit board industry to produce LCD systems.

By 1993, with technological sophistication and an enhanced company reputation, non-OEM accounted for a majority of sales (85 per cent or so), but the company still valued OEM as a useful (if relatively minor) source of technology acquisition. In 1993 the company employed around 30 qualified engineers and technicians, including several of the original founders. Among its customers were Mercedes Benz (of Germany) and GEC (of the UK).

One of the firm's most successful projects was a proprietary design of a touch-sensitive, portable hand held terminal, which was patented worldwide. Around 40,000 units were first sold to the Royal Hong Kong Jockey Club to enable customers to place bets automatically. A new project aimed to apply the hand-held system to other applications in inventory control, point of sales, instrumentation and banking. The firm supplied many high-technology niches and could still respond to orders as small as US$1,000 in 1993.

Combining Latecomer and Leader Advantages

Benefiting from US training and from Hong Kong's university facilities and faculty, Varitronix very quickly began to innovate with new products. It simultaneously combined its own innovative capabilities with the latecomer advantages of other Hong Kong firms, notably the low-cost Chinese and Malaysian production facilities. By coupling latecomer with leadership advantages, Varitronix succeeded in a range of high-technology niche markets. Like many others, Varitronix exhibited a mixture of leader, follower and latecomer features, retaining some of the latter (e.g. OEM) to its advantage.

The firm grew up with the LCD industry and kept abreast of the technology. By 1993 it spent around 6 to 10 per cent of annual turnover on R&D, in line with market leaders and ahead of most of Hong Kong's more traditional

OEM-oriented latecomers. As with other Hong Kong firms, most of the design and development work was carried out in the headquarters in Hong Kong, while manufacturing engineering was carried out in China and Malaysia.

7.12   HONG KONG'S TECHNOLOGICAL POTENTIAL

Hong Kong, like the other three NIEs, broadly conformed to the model of Chapter 3. However, despite its success, Hong Kong lagged behind the others in technological capability. In 1990 Hong Kong electronics exports were only one-half of Singapore's. Although it is impossible to judge how the industry would have developed under alternative policy regimes, R&D spending was low on average, the largest firms were smaller than their counterparts in Taiwan and electronics had yet to catch up with clothing as the largest export sector.

Most firms still focused heavily on consumer electronics and the industry had yet to make a substantial transition to industrial systems. In semiconductors, the core activity of wafer fabrication had not yet taken root in Hong Kong as it had in the other dragons.[13] This chapter therefore agrees with other studies which argue that Hong Kong's electronics potential has yet to be fully realized (e.g. HKPC 1982; IEEE 1991 p. 56).

Nevertheless, the industry should be proud of its achievements in electronics. Latecomer firms pursued bold strategies of export-led technology learning, similar to those in the other NIEs. Companies linked their innovative efforts to export market needs and overcame the disadvantage of the small local market. As firms approached the technology frontier, channels of learning evolved from OEM and sub-contracting to ODM and OBM, albeit in less complex goods than the other three dragons. While most firms were too small to break free of heavy dependence on sub-contracting, the example of Varitronix showed how high-technology start-ups were able quickly to combine the cost advantages of being a latecomer with the leadership benefits of being a technological innovator.

The chapter also showed how the latecomer firms contributed to the rapid economic growth of Mainland China, providing access to international markets and technological know-how. Hong Kong firms, like their Taiwanese counterparts, assisted China to overcome its isolation and to follow the NIEs in a fast growth path. The combination of Mainland China's low-cost labour and engineering with latecomer know-how and finance provided a powerful additional force to Pacific Asia's regional economic progress.

NOTES

1. Interviews were conducted with 20 or so Hong Kong electronics firms in 1993, including six of the largest local companies and two TNCs. Data were also collected from universities, administration offices and other institutes connected to electronics.
2. As Section 7.7 shows, many of these firms would later return to Guangdong bringing an electronics export industry to China. Much of China's FDI in electronics is either from Hong Kong or routed through Hong Kong from Taiwan.
3. See the case study of Varitronix (Section 7.11 below).
4. Total electronics employment in Hong Kong stood at around 85,000 in 1990. Many companies also employed large numbers in China.
5. Calculated from IEEE (1991 p. 51).
6. Calculations based on Fok (1991 pp. 263–4).
7. According to company interviews, Taiwan began large-scale outsourcing to China in 1987 and 1988, four or five years later than most Hong Kong firms.
8. Motorola interview (Hong Kong, 1993). This confirms the earlier data in Thoburn et al. (1991 p. 46).
9. One company study estimated that in 1992 the cost of a software engineer in the US was around US$80,000 on average, compared with US$10,000 in China. The figures include overheads and other additions. The proportion of wages in the China case was very low indeed.
10. As noted earlier, TI worked in partnership with RJP, Anam and other firms to transfer technology in order to increase their sales in the region.
11. Although Motorola competed head on with Japanese companies such as Sharp and Sony in some areas of consumer electronics, most key components for consumer goods (e.g. TVs) were supplied to latecomers by Japanese rather than US firms.
12. LCDs are electronic devices made up of two pieces of glass with a fluid (liquid crystal) in between. The inside of the glass is coated with a conductive, patterned film of electrodes which respond to electric signals by forming alphanumeric characters or graphical patterns.
13. At the time of the research three wafer fabrication facilities were identified in Hong Kong (Elcap, Wah Fong and one other). Site visits by the author confirmed that the facilities were small and lagged behind those in the other dragons by about three to four years.

8. Conclusions and implications

8.1 INTERPRETING THE FINDINGS

This book has shown how East Asian latecomer companies built up their formidable competitive capabilities and successfully learned the technology of electronics. As a result of their strategies, economic growth, manufacturing, trade and innovative capacity, have progressively shifted to Pacific Asia from Europe and the US.

This final chapter brings together the themes of the book, summarizing the main findings and drawing implications for theory and policy. A simple framework is introduced to analyse the technology strategies of latecomer firms, to show how they learned to innovate and catch up with international market leaders. The chapter assesses the resulting structural orientations, strengths and weaknesses of the latecomers and predicts the strategies they are likely to adopt to become full competitors on the international stage.

The East Asian latecomer experience has important implications for innovation theory and the currently popular debate concerning the causes of the Pacific Asian miracle. By comparing the experience of the four NIEs, the chapter is able to emphasize the range and diversity of East Asian models of industrial development and identify some of the common principles which underlie the achievements of the Asian dragons.[1]

Although the challenge to Japan is still very much in its embryonic stage, there are indications that the four dragons have acquired significant innovative capabilities to overcome most of their early latecomer weaknesses. In partnership with foreign TNCs, the dragons have begun to widen their scope of operations into China and the second-tier NIEs, overtaking Japan as the largest investor in the region. Finally, the chapter asks whether there are any lessons for other countries, developing and developed, wishing to learn from and respond to the East Asian latecomers.

8.2 LIMITS TO THE ELECTRONICS CASE

In drawing conclusions it is important to recognize the special characteristics of the electronics industry. Even though it is the largest, most competitive export industry in East Asia it has its limits. The industry is of a particular kind and generalizations need to be drawn very carefully. It is an example of a fast-growing, internationally traded industry in which the division of tasks across national boundaries is technologically possible and advantageous to TNCs. Electronics is also a manufacturing-driven, high-throughput industry where the cost of workers, technicians and engineers plays a crucial part in competitive advantage.

Not all manufacturing industries have these characteristics. For instance, the relevance of electronics to customized, complex systems industries such as intelligent buildings, nuclear power equipment, helicopters, large aircraft, flight simulators and many large-scale capital goods is far less clear.[2] Competitive advantage in these industries is seldom, if ever, centred on volume production costs and incremental process improvements as in electronics. Complex systems industries are small-batch, project-based, design-intensive industries where advanced software expertise and continuous interaction with industrial users is essential to competitive success. In these areas, Pacific Asia, including Japan, may lag behind Europe and the US (Miller et al. 1995). In the future, entry into these industries will become an important competitive challenge for the latecomers.

However, electronics is the largest East Asian export sector and much can be learned from the cross-country comparison. The lessons from electronics may well apply to other fast-growing export industries including bicycles, clothing, athletic footwear, and sewing machines. As Chapter 5 showed, in several of these sectors innovation capacity had also shifted to Pacific Asia, while Western manufacturing had declined. As well as being the leading industry in the region, electronics has also had a wide demonstration effect, showing to firms and policy makers what can be done. It has led directly to the development of a variety of important support industries which have grown up alongside electronics. Therefore despite its limitations, it is of vital interest to the region.

8.3 LATECOMER LEARNING STRATEGIES

Chapter 3 argued that in order to catch up East Asian firms would have to overcome two sets of extreme latecomer disadvantage. First, they would have to surmount their dislocation from the main international sources of innovation, technology and research. Second, they would need to overcome their distance from the advanced markets and user-producer links essential to innovation. Only by overcoming these latecomer difficulties could firms achieve export success.

The book has shown that a variety of learning strategies enabled the latecomers to overcome their difficulties and considerably narrow the tech-

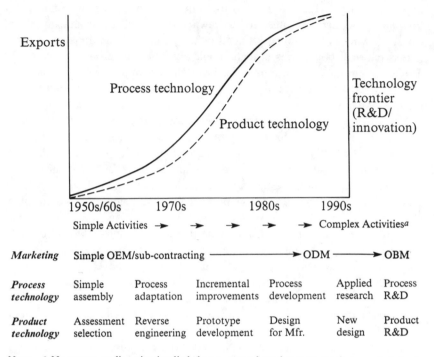

Note: [a] No stages or linearity implied, but a general tendency to catch up cumulatively, through time with capabilities building systematically upon each other.

Figure 8.1 Latecomer firms – export-led learning from behind the technology frontier

nology gap with the West and Japan. Figure 8.1 presents a simple framework for analysing the nature, direction and historical determinants of learning in order to encapsulate the evidence from each of the country chapters. As the model suggests, latecomer firms engaged in a process of export-led technological learning. The left-hand vertical axis represents electronics exports, which also correspond to employment and output growth. The right-hand vertical axis represents the innovation frontier, defined as the point at which R&D becomes central to competitiveness.[3]

The frontier is a constantly moving, dynamic target. In order to catch up firms must constantly narrow the gap between themselves and the market leaders. The horizontal axis represents the acquisition of process and product technologies by latecomer firms through time, beginning with simple activities such as assembly and graduating towards more complex tasks such as process adaptation and R&D. The relation between export growth and technological acquisition is presented in the shape of a diffusion curve, which corresponds to the slow initial start-up of the 1950s and 1960s, the adoption phase of the 1970s and the rapid take-off in the 1980s. As more latecomers entered to profit from new opportunties, others followed in a Schumpeterian swarming fashion, bringing about the surge in growth during the 1980s. The curve also suggests the possibility of a slowdown in the future as more sub-sectors of electronics mature.[4]

As the model suggests, latecomers passed through various historical stages of technological development. The first latecomer firms emerged in the 1960s, producing labour-intensive products under joint ventures or sub-contracting arrangements with Japanese, US and European firms. TNCs and foreign buyers were initially attracted to East Asia by low labour costs. Foreign firms supplied training, advice on manufacturing processes and product styling, as well as capital goods. At the same time, local technicians, engineers and managers were trained within the subsidiaries of the TNCs. The larger foreign buyers and TNCs supplied formal training courses for assembly workers and technicians to ensure that quality and delivery targets were met. Some TNCs worked closely with local sub-contractors to buy low-cost parts and components, giving rise to a variety of electronics support industries.

Through the 1960s and 1970s the latecomers learned by manufacturing simple consumer electronics and by assembling and testing semiconductors. Firms entered from the electrical goods sector, clothing and other industries. Others companies started from scratch. Some individuals left their jobs in TNC subsidiaries and began their own companies, often supplying their former employers. Gradually the latecomers learned by upgrading their production processes and by efforts to improve the quality and speed of manufacturing.

After a long period of learning the art of assembly, during the 1980s electronics took off, overtaking clothing and other sectors as the largest export industry in each country except for Hong Kong.[5] To meet more complex export demands, latecomer firms were forced to acquire additional technological know-how and skills. Some latecomers learned to design products independently of the foreign buyers, although most continued to manufacture for the TNCs under OEM arrangements. New entrants benefited from the improving technological, communications and transport infrastructures. They also benefited from the growing support industries and, above all, the supply of well-educated human resources. Some firms learned by imitation. Others, such as ACER of Taiwan, entered as engineering-intensive, small start-ups with advanced capabilities.

The first latecomers (e.g. Anam of South Korea) tended to graduate from simple to advanced learning over a period of two decades or more. Later entrants (e.g. MTI of Taiwan) began at more advanced stages, missing out early phases. Learning by imitation took place once foreign TNCs had demonstrated the advantages of local production. Local firms learned under OEM and sub-contracting arrangements. Learning-by-OEM was a training school by which many local firms entered the industry, secured a market channel and acquired technology.

Many of the later entrants very quickly advanced to innovative learning, developing their own designs and competing, at least in some areas, at the technological frontier. However, as the case studies show, even the most advanced firms still operated as latecomers in the early 1990s. Many still produced a large proportion of their exports under sub-contracting and OEM arrangements, depending on foreign companies for access to markets, key components and capital goods.

Learning occurred within TNC subsidiaries, especially in Singapore where foreign firms dominated. The TNCs became more technologically advanced as time progressed, narrowing the gap between themselves and their parent plants abroad. They benefited from East Asia's rapidly developing technological and market base, especially during the 1980s. As a result of Pacific Asian investments, some European and American TNCs have rejuvenated their operations and learned from the East Asian context.

Investing in skills enabled latecomers to increase their export earnings by broadening their customer base and expanding their product ranges. Learning to master production know-how and to design new products enabled them to capture a larger share of the value-added. During the latter half of the 1980s, each of the four countries made the transition from consumer electronics and simple assembly activities to complex industrial systems, including computers, semiconductors, disk drives and peripherals.[6]

By the late 1980s industrial electronics represented the largest electronics export in Pacific Asia. Into the 1990s, the latecomers continued to use their entrepreneurial skills to exploit foreign technology and market channel to their advantage. Gradually, the leading latecomers (such as Samsung and Anam of South Korea) narrowed the technology gap with the market leaders and were able to negotiate more equal partnership with Japanese, European and US TNCs.

The evidence showed that latecomers pursued tenacious and bold strategies towards technological acquisition and international marketing. Through training, hiring and learning, firms transformed their initial low-cost labour advantages into highly competitive low-cost precision engineering and management. Some acquired foreign firms in Silicon Valley and other locations to gain technological skills and access to markets. Others formed long-term R&D partnerships with foreign market leaders.

In some cases, experimental design and R&D began fairly early on and there was feedback between the early and later stages.[7] However, as the evidence shows, there was a general tendency for firms to begin with simple tasks and accumulate capabilities systematically in a path-dependent, cumulative manner, with skills and knowledge gradually building on each other.

As the S-shaped curve in Figure 8.1 suggests, rapid growth is likely to be followed by a slowdown. Eventually, as in the case of consumer electronics, older sectors mature and growth decelerates. However, the rate of slowdown need not follow any particular path. Specific technological bottlenecks can be overcome. As wages rise, local and regional markets may begin to absorb a larger share of total production, as has already occurred in South Korea and Taiwan.

Government learning also progressed in the four dragons. States encouraged firms to expand their electronics activities by improving the educational and technological infrastructure. Infrastructural provision was not a once-and-for-all activity. On the contrary, it evolved and continued to evolve according to the needs of the economy.

To sum up, latecomer firms systematically acquired foreign technology from others through their learning strategies. Through these efforts, the electronics industry grew rapidly as did East Asia's innovative capabilities.

8.4 LINKS BETWEEN TECHNOLOGY AND THE EXPORT MARKET

As the evidence shows, learning extended beyond technology to marketing. Firms learned to package, distribute and market their goods. Some established marketing departments at home and then in the advanced countries. Marketing know-how enabled firms to diversify their customer base and to increase their growth opportunities. Like technology, export marketing involved substantial investments in skills and organization. Ultimately, the larger latecomers established their own brand names abroad, organized their own distribution outlets and advertised directly to customers. However, by the early 1990s only the very largest South Korean and Taiwanese firms had established well-known brand names.

There were many connections between the stages of technology and market development. Firms learned to improve both their technology and marketing skills simultaneously in order to increase profit and market share. The channels for learning technological and marketing capabilities were often one and the same, as with the OEM/ODM system. To increase sales of production capacity to key customers, joint engineering projects were carried out, enabling latecomers to share the costs of learning with their customers. Later on, to bring innovative new products to the market, the larger firms made heavy, long-term investments in R&D.

As the country chapters showed, exports acted to pull forward latecomer skills and know-how, enabling firms to overcome the lack of demanding buyers and user–producer links enjoyed by leaders and followers. By using exports as a discipline to lead technological development, companies overcame the deficiencies of local markets. Exports provided the missing demand-pull mechanism allowing manufacturers to narrow the innovation gap. Local competition stimulated innovation, successful exporters were imitated by others, and gradually the relocation of production to the NIEs occurred.

8.5   THE OEM SYSTEM

Through OEM, sub-contracting and other channels, export demand provided the focusing device for learning and forced the pace of progress. Over the past 30 years or so, latecomers developed and exploited a variety of institutional mechanisms for exporting overseas, monitoring international markets and guiding local investments in technology. In electronics, none was more important than the OEM system. Under OEM, the TNCs purchased large quantities of goods manufactured by the latecomers. By selling the products under well-known foreign brand names the latecomer avoided the need for heavy investments in marketing and distribution.

Under the early OEM deals, the foreign corporations frequently supplied training, technical specifications and advice on engineering and capital goods. The OEM system proved an enduring technological training school for latecomers in the NIEs, enabling hundreds of small firms to overcome barriers to entry. By the early 1990s a significant proportion of electronics output was still sold under OEM and similar sub-contracting arrangements.

8.6   FROM OEM TO ODM

During the 1980s, the OEM system was developed, refined and continuously re-thought. With advances in latecomer design capabilities, by the late 1980s foreign buyers and TNCs had begun purchasing goods under so-called ODM, allowing local companies to exploit their design talents and thereby gain more of the added value. Sometimes the latecomers designed goods independently, using their own knowledge of the international market. In other cases they worked closely with foreign buyers and TNCs. The emergence of ODM signified a new phase of latecomer technological progress, indicating that local firms had internalized much of the ability to understand market needs, then to design, develop and make electronic products for overseas markets.

As with OEM, the ODM system allows the foreign buyer or TNC to brand and distribute the goods manufactured locally, enabling the latecomer to circumvent the need for heavy marketing investments. The South Korean chapter showed that, although the term ODM was not in common use as it was in Taiwan, the leading *chaebol* had reached an equivalent technological stage, designing their own products in anticipation of international design needs or to general requests from foreign partners. In the Taiwanese computer industry, ODM was encouraged by large international trade shows which attracted US firms such as IBM, Apple, Dell, Intel, AT&T, and Japanese firms such as Mitsui and Hitachi. Taiwanese PCs and notebook computers were purchased in large volumes and sold in discount stores and other outlets in Japan and the US, usually bearing the brand of the TNC buyer. In computers alone, Taiwanese ODM sales reached around US\$3.5 billion in 1993, around half of Taiwan's total computer exports.[8]

The book showed how ODM evolved out of the OEM system, highlighting the growing inter-dependence between latecomers and TNCs, and the addiction of many leading companies to the manufacturing skills of the latecomers. Today, it is doubtful if many of the largest international computer suppliers could survive without the hundreds of low-cost Taiwanese suppliers of PCs and peripherals and their low-cost operations in China. Still, most of the ODM suppliers, including medium-sized companies such as Inventec, Datatech, Compal, First International and Twinhead, are virtually unknown in the West.

8.7   THE NATURE OF LATECOMER INNOVATION

A dynamic, unique and powerful system of innovation has gathered pace in East Asia. The case of electronics indicates how far that system has come and suggests that it will go much further as the creative design talents of latecomer firms are released. Through the 1980s innovation was primarily driven by the need to master manufacturing technology. In the 1990s and beyond, product design, development and research will increasingly take centre stage.

Innovation in East Asia cannot be measured solely by counting individual process and product changes, although many were documented in the study. Ultimately the cumulative sum of the minor, incremental improvements to processes and products will result in more dramatic new innovations. But, so far, these have yet to dominate the Pacific Asian scene.

The country chapters showed that latecomers progressively built up innovative capabilities in order to improve their market opportunities. Patterns of innovation were conditioned by the origin, advantages and disadvantages of latecomer industrialization. The core of much of the innovation was continuous improvements to manufacturing processes. Early benefits included the lowering of production costs. Later, firms added productivity gains to technician work, engineering and management. Once process capabilities were in place, firms progressed to product innovations, often working alongside their foreign partners. Most recently there were signs of significant new product designs from the leading firms.

In contrast with Western preoccupations with invention, R&D and advanced product designs, East Asian innovation developed out of the competition to manufacture goods for established markets.[9] The focus of Western innovation literature on R&D and product design reflects its concern with leaders and followers. However, East Asian latecomer innovation grew out of the need to compete from behind the technology frontier and their ambition to catch up.

The comparison with standard innovation models is striking.[10] In contrast with normal Western models, East Asia's latecomers travelled backwards along the product life cycle, reversing the normal path. Firms began by learning mature, standardized manufacturing processes. Once manufacturing capabilities were in place, companies moved onto advanced process engineering, product–process interfacing and product design. Only recently and selectively have the leaders exploited R&D for future product developments. In this sense, they reversed the normal cycle of innovation, passing from mature to early stages of the product life cycle, from standard to experimental manufacturing processes and from incremental production changes to R&D.

Innovation in East Asia makes no emotional or ideological distinction between innovation and imitation. The patterns of imitation demonstrate corporate creativity and result in competitive advantage, bringing about industrial transformation and development. Learning to imitate allowed latecomers to expand their range of customers and to improve their production capabilities. Imitation led to pioneering, creative product innovation, as it did in Japan.[11] This is the natural sequence of latecomer technological development.

By focusing on manufacturing processes, latecomer firms galvanized and exploited their key resources and advantages. By the early 1990s many East Asian firms operated at the early stage of the product life cycle, designing their own electronic goods, improving the interface between design and production and relying more on in-house R&D. Their innovation path enabled them to exploit their cost advantages and circumvent direct competition with market leaders. Dedication to manufacturing resulted in an unparalleled ability to respond to fast changing electronics demand throughout the world.

Some of the most important innovations were organizational. Anam of South Korea, by 1993 a US$2 billion corporation, pioneered the idea of sub-contract semiconductor packaging and became the largest firm in the world in this market. TSMC of Taiwan was the first company worldwide to carry out fabrication-only services in volume for chip designers. Other companies followed strategies of combining low-cost sub-contract work with highly innovative product designs for niche markets.

The most important organizational innovations were the new forms of vertical inter-company partnerships formed under OEM and ODM. These enabled hundreds of firms to exploit foreign export channels, to overcome barriers to entry and to systematically acquire technology. They allowed the latecomers to couple the technology pull of Western and Japanese markets with local technological efforts. The expansion and sophistication of OEM, sub-contracting and licensing provided the route for technology transfer from advanced users in the West and Japan, forced continuous improvements upon local firms and enabled the latecomers to sub-contract their way up the technology ladder.

The form, depth and breadth of East Asian organizational innovations are new to the marketplace and to the world and therefore constitute innovation in the most meaningful sense of the word. Indeed, the manner in which regional development occurred under OEM and sub-contracting with foreign firms is a significant departure with no obvious historical parallel. It is a new large-scale feature of economic development only witnessed in the latter part of the 20th century.

8.8   DIVERSE POLICIES MODELS

While the book focused primarily on firms, it would be wrong to suggest that latecomer companies developed in an economic or policy vacuum. Among the four NIEs there was enough variety to make useful comparisons between government policies, industrial structures and patterns of ownership in electronics. These differences reveal important insights into the underpinnings of East Asian economic development.

As Chapter 2 pointed out, the Japanese-oriented flying geese model of East Asian development fails to account for several important forces in the region's progress, in particular the distinctive development styles of the four dragons. These include not only the overseas Chinese character of Taiwan, Singapore and Hong Kong but also the aggressive efforts of the South Korean *chaebol* to catch up.

Figure 8.2 summarizes the diversity of development approaches in electronics.[12] One striking policy contrast was the degree of government intervention in electronics. Both South Korea and Singapore followed highly interventionist policies. Singapore intervened mainly indirectly through

High	**Korea** Large firms Locally owned	**Singapore** Large firms Foreign owned
Low	**Taiwan** Small local firms Large foreign firms	**Hong Kong** Small local firms Large foreign firms

Government policy: degree of direct intervention *(vertical axis label)*

Low/closed   **High/open**

Degree of openness to
FDI and foreign imports

Figure 8.2   Diversity of policy models: the four dragons

subsidies and other inducements to TNCs as well as infrastructural and educational policies, often for the benefit of specific foreign firms. South Korea intervened both indirectly and directly in the strategic affairs of the *chaebol*, offering cheap finance, setting export targets, preventing some diversifications and allowing others.[13]

In stark contrast to South Korea and Singapore, Hong Kong pursued a non-interventionist, *laissez-faire* approach to electronics and economic development in general. In the case of Taiwan, the government intervened selectively in scale-intensive areas such as semiconductors, but left most export activity to the strategies of private companies in the market place.

Figure 8.2 also illustrates important differences in orientation of industrial policy. While Hong Kong and Singapore pursued strictly conventional export-led policies, South Korea and Taiwan combined these policies with import substitution, controlling or banning imports to protect local firms and using government procurement to stimulate local enterprise. South Korea was the most restrictive, receiving much less FDI than the two city states, despite its greater size. Taiwan often negotiated the terms of FDI and tied TNCs to local content rules and export targets. In sharp contrast, Singapore and Hong Kong encouraged FDI with low taxation, special incentives and welcoming policies and schemes, allowing a degree of freedom seldom witnessed in South Korea and Taiwan.[14]

Regarding company size, while Taiwan and Hong Kong depended to a large extent on small, Chinese family businesses, the South Korean Government patronized the very large conglomerates, which today rank among the largest in the world. South Korean policies resulted in highly concentrated industrial structures, with the *chaebol* dominating electronics and many other industries. By contrast, in Hong Kong and Taiwan small firms proliferated, resulting in a highly dispersed industrial structure in electronics. When comparing industrial concentration, it is difficult to imagine a more stark contrast between South Korea and Taiwan. An important lesson from Taiwan and Hong Kong for other developing countries is that small scale need not necessarily be a hindrance to export growth in sectors such as electronics.[15]

Policy and company strategy were closely entwined. Small size led overseas Chinese firms to rely on speed and flexibility, while the large South Korean companies took a high-volume, process-intensive approach to electronics. Many Taiwanese and Hong Kong firms specialized in fast-changing market niches. South Korean policies and corporate strategies owed much to the Japanese *keiretsu* which provided a nearby role model.[16] By contrast, Taiwan's approach drew from a variety of sources. Local firms combined their traditional overseas Chinese business styles with modern management training received in leading US corporations, universities and business schools. Many of Taiwan's high-technology firms owe more to the American management influence than to the Japanese.

Regarding ownership, Taiwan and South Korea relied mostly on locally owned firms, while Singapore depended almost entirely on foreign TNCs. TNC investments were encouraged because the government believed, rightly or wrongly, that the local entrepreneurial base was too weak to lead industrialization.[17] Successive policies encouraged TNCs to transfer technology to the subsidiaries and to build up Singapore as a leading South East Asian centre for electronics manufacturing.

In short, the evidence shows striking contrasts between the four non-Japanese models of East Asian development. Policy diversity led to plurality in industrial concentration, corporate ownership and strategy, patterns of innovation and paths of industrial development. Despite these differences, each of the four NIEs achieved unprecedented economic and technological progress. Therefore, it is necessary to look to the common factors to explain each country's success.

8.9 PRINCIPLES OF SUCCESS

At least four important similarities in macroeconomic policy, industrial orientation and technological development explain how each country succeeded in electronics and overcame the latecomer disadvantages outlined in Chapter 3.

First, firms undoubtedly benefited from low rates of interest, relatively low inflation and high savings. There can be no doubt that achieving macroeconomic stability by getting the basics right was a key factor in East Asian development. Stability, coupled with the commitment of each government to industrial development, provided firms, groups of firms and industry associations with an environment for long-term planning and investment.

Second, the latecomers responded to the outward-looking, export-led industrial policies of each country. Export-led growth provided the framework to enable firms to overcome their dislocation from the centres of world innovation and demanding international markets, providing the demand-pull for innovation in East Asia. Where import-substitution was evident, as in the case of South Korea and Taiwan, import restrictions were conducted within an overall policy of export-led growth. Exports acted as a focusing device for technology investments and encouraged the growth of a variety of institutions to enable exports to flourish. Arrangements such as OEM, joint ventures, licensing and sub-contracting were encouraged by government policies, allowing firms to acquire and adapt foreign technologies.

Third, each NIE developed an appropriate educational and technological infrastructure. In the early stages removing illiteracy and supplying a sound general education was important for industrial development. Once literate, children then went on to receive vocational education in the crafts and in engineering. By developing, adapting and improving training and education policies, each country supplied a sufficient number of technicians and engineers for firms to utilize. Each country set up institutes for engineering training and support for industry; many firms benefited from their services and supplies of well-trained engineers and technicians. Vocational courses, often directed towards company needs, were carried out in local universities and polytechnics.

Fourth, where and when necessary, governments intervened to ensure that the entrepreneurial base was strong enough to lead industrialization. Without a sufficiently talented cadre of firms, no industrial strategy can be successful almost by definition. Policies to overcome what can be called entrepreneurial (or corporate) failure took various forms.[18] In Singapore in the 1960s the quality and quantity of local firms was judged by government to be inadequate to lead industrialization. It therefore set about attracting TNCs to develop the electronics industry and took control of other industries itself. In South Korea market mechanisms and institutions were also inadequate. The domestic government built up the large *chaebol* to overcome the problem of corporate failure. In Taiwan, in many scale-intensive sectors state-owned firms were established to organize industrial development.[19]

8.10 IMPLICATIONS FOR DEVELOPING COUNTRIES

When drawing implications for other developing countries, it is important to remember that each economy starts from its own unique set of advantages, disadvantages, opportunities, sectoral strengths and weaknesses. There can be no direct transfer of East Asia's models to other countries, especially in electronics. During the 1990s, opportunities in electronics will increasingly be exploited by China and the second-tier NIEs of East Asia. Other developing regions will face more intense competition from East Asia and, the likelihood is, growth in several of the electronics sub-sectors will slow as markets mature. However, the principles identified above do present important general considerations of value.

The East Asian experience emphasizes the importance of macroeconomic stability in providing an environment conducive to long-term corporate planning and investment. In order to fully exploit technological opportunities and to optimize any form of industrial development, policies towards inflation, interest rates, balance of payments and other macroeconomic essentials have to be implemented effectively. It is difficult to see how countries with accelerating inflation and constant industrial turbulence and instability can achieve sustained, long-term industrial development. Similarly, outward-looking, export-led policies are essential to stimulate innovation. Policies which focus primarily on supplying local markets cannot achieve the export-led technology pull exercised by export markets. Only by supplying the most demanding users can the latecomer market dislocation factor be overcome.

For some countries the lesson from East Asia is that a dynamic entrepreneurial base needs to be established to exploit new market opportunities and to acquire technology. This corporate base cannot be taken for granted as it is in many studies of the East Asian development.[20] In some countries, policies to stimulate the development of firms and to overcome corporate failure may be necessary. The entrepreneurial base may involve locally owned firms, joint ventures and foreign subsidiaries. As the book showed, if advantageous to them, TNCs will transfer substantial amounts of technology both to their subsidiaries and to local firms in developing countries. TNCs trained local engineers, formed educational links within East Asia and provided an important channel into overseas markets. However, as stressed throughout the book, it was the actions and strategies of local companies which ensured that TNCs' investments were exploited to maximum effect.

Finally, the East Asian educational and infrastructural experience is instructive. The removal of illiteracy and the supply of vocational education was an essential feature of overall industrial development. In electronics local firms demanded recruits with a range of basic engineering and industrial skills. The implication for other countries is that educational policy needs to account for the continuously changing needs of industry. Policies should account not only for the high-technology element of new industries, but all the basic technician, craft and engineering skills needed to support those industries.

8.11   IMPLICATIONS FOR LEAPFROGGING

Rather than leapfrogging from one vintage of technology to another, the study (and Chapter 6 especially) showed that TNCs and local East Asian firms engaged in a painstaking and cumulative process of technological learning: a hard slog rather than a leapfrog. The route to software and advanced information technology was through a long difficult learning process, driven by the manufacture of electronic goods for export.

Even at today's advanced stage, the competitive advantage of East Asia's latecomers is low-cost, high-quality production engineering, rather than software or R&D. Although the NIEs are increasing their investments in science and advanced technology, they remain conspicuously weak compared with Japan and other OECD countries.

Also in contrast with leapfrogging, much latecomer learning took place in a field which could be described as pre-electronic: mechanical, electromechanical and precision engineering activities, for example. Competencies tended to build upon each other incrementally, leading to advanced engineering and software. Firms tended to enter at the mature, well-established phase of the product life cycle, rather than at the early stage, again contrary to the leapfrogging idea.

The policy implication of this finding is that to build an electronics industry, local firms require human resources trained in a range of basic craft, technician, engineering and industrial skills, rather than the software and computer-based skills normally associated with information technology. Like the NIEs, other developing countries should take very seriously the low-technology side of so-called high-technology industries. Only by developing capabilities in fields such as plastics, mouldings, machinery, assembly and electromechanical interfacing, did East Asia emerge as the leading export region for electronics.

8.12   THE EAST ASIAN MIRACLE DEBATE

The book went beyond the long-standing argument over whether state policies or market mechanisms best explain East Asian development. The market *versus* state debate failed to provide any insights into the nature, costs, mechanisms and processes of innovation in East Asia. Indeed, the traditional literature has had very little to say about the strategies by which latecomer firms acquired technology and succeeded in international markets. This book has shown how firms learned to export from behind the technology frontier by bold, deliberate and creative strategies. No amount of analysis of government policies, general economic data or abstract market mechanisms could show how technological learning occurred in Pacific Asia or how the region expanded its global export reach.

8.13   THE CHALLENGE TO JAPAN

At the present time the challenge to Japan from within East Asia is still at an embryonic stage. Japan is still overwhelmingly dominant both in economic and technological terms. Japanese companies out-sell and out-invest other firms and their capabilities and resilience should not be underestimated. As the study has shown, each of the NIEs is dependent on Japan for capital goods, key components and new product designs in electronics. Many latecomer firms remain strategically subordinated to Japanese companies under OEM, ODM and other sub-contracting arrangements.

However, Japan's deep recession of the early 1990s shows that the economy is not invulnerable, nor are Japanese firms. Japan's market shares in computers, semiconductors, electronics capital equipment and automobiles began to decline as US firms restructured, the Yen appreciated and low cost East Asian competitors surged forward.[21] In 1992, for the first time in several decades, two of the largest Japanese electronics producers reported heavy losses. All six Japanese leaders reported declining sales and falling profits.[22] During the 1980s, Japanese electronics output grew by a compound rate of around 10 per cent per annum. In 1992, sales shrank by around 10.6 per cent to US$189 billion. In the key area of semiconductors, the US overtook Japan as the world's largest supplier of traded chips, partly because of the success of South Korean firms in gaining market shares in the Japanese stronghold of commodity DRAM chips. It also became clear that Japanese firms had failed to make much headway into complex semiconductors such as microprocessors, advanced software and computing technology. To some, Japan's problems were more than just cyclical, amounting to a deep structural malaise. Japanese firms, like IBM of the US, appeared too large, bureaucratic, slow and conservative, according to many observers.[23]

Such reports may underestimate Japan's corporate resilience and ability to restructure. However, as this book has shown, an irreversible surge of innovative capacity among non-Japanese East Asian firms occurred during the 1980s, re-casting the dominant position of Japanese firms and challenging

Japan as the single dominant economic power in the region. The four NIEs rapidly progressed from simple goods to disk drives, semiconductors, computers and other high-technology industrial goods, creating a new challenge to Japan from within Asia Pacific. Today, in many of Japan's traditional industrial strongholds competition is felt most strongly from other East Asian rivals, rather than from the US or Europe. Recognizing this as a long-term opportunity, many leading US and European firms have invested heavily in the region, forming strategic partnerships with latecomer companies in an effort to capture market share and benefit from the climate of innovation in non-Japan East Asia. Japanese firms have also invested heavily in the lower-cost countries of the region, stimulating the high-technology manufacturing infrastructure outside of Japan and encouraging further innovation in non-Japan East Asia.

The four dragons will continue to be major outward investors in East Asia, particularly in the fast-growing markets of Malaysia and China. The emergence of China as a large, fast-growing market poses problems as well as opportunities for Japan. Chinese growth is likely to continue to rely heavily on the return of overseas Chinese capital. Partnerships between the Mainland Chinese firms with Taiwanese and Hong Kong latecomers are likely to present significant low-cost competition to Japan and could lead to the emergence of China as a significant technological power over the next decade.

Largely excluded from the markets of Japan, many leading American and European firms have been welcomed as investors in Hong Kong, South East Asia and China. Many examples of how such alliances have benefited both US, European and latecomer firms were illustrated in this book. Some US and European TNCs have grown rapidly in East Asia, revitalizing their operations and learning new skills. In alliance with the latecomers, these TNCs present a new dynamic within the region. Only time will tell if and how the competitive pecking order will change. However, there can be little doubt that innovation in the NIEs has permanently altered the competitive dynamics of East Asia and the nature of the challenge to Japan.

8.14   FACING THE INNOVATION FRONTIER

The origins, paths and strategies of latecomer firms explain their current advantages, weaknesses and opportunities. Their solution to overcoming barriers to entry was to couple technology development to export market needs under sub-contracting arrangements, the OEM system, licensing and other institutional mechanisms. This enabled them to learn from foreign companies and to rapidly expand their exports. As they climbed the technological ladder, they transferred out labour-intensive activities to Malaysia, Indonesia, Thailand and the South East coastal region of China. This resulted in deeper Pacific Asian economic integration and a burgeoning of intra-regional trade.

The strategies of East Asia's latecomers also pinpointed their continuing weaknesses. Only in a small number of areas have major new product innovations been generated by local firms. In most fields, they are still dependent on their natural competitors for key components, capital goods and distribution channels. Latecomer firms suffer from weak R&D capabilities and poor brand images abroad. Without stronger product innovation capabilities they will continue to rely on a mixture of catch-up, imitation-based growth and incremental innovation. Lacking strong R&D capabilities and a thriving capital goods sector in electronics, the technological roots of the latecomers remain shallow.

The majority of latecomers are still distinct from followers and leaders. Although some have made the transition to follower and leader in some areas, most are highly dependent on OEM and sub-contracting for access to markets and technology. As more latecomers approach the innovation frontier, they will require new strategies to gain technology and to overcome remaining weaknesses. Many will adopt the strategies of followers and leaders by increasing their R&D expenditures and improving their brand images abroad. Others will gain advantage by expanding their basic OEM activities into neighbouring low-cost areas of East Asia. However, most are likely to pursue a mixture of both strategies, creating hybrid leader/follower/latecomer corporations able to benefit from both leader and latecomer advantages and able to control and contain latecomer disadvantages. Such hybrid strategies, already in evidence in this book, will prove a powerful competitive formula in the future.

To continue their success, much will depend on whether East Asian firms can overcome their latecomer disadvantages in design, capital goods, R&D and marketing. The strategy of larger companies is to invest heavily in R&D and in brand awareness campaigns. However, the results of these strategies are still unfolding and so far the results are mixed. As shown in the case studies, some firms retreated back into OEM/ODM in the early 1990s after sustaining heavy losses in own-brand investments. Others witnessed the cost and risk of directly challenging the US and Japanese leaders, choosing to rein back some of their ambitions for the time being. Of vital interest in the 1990s and beyond is whether, and to what extent, the latecomers are able to shed their remaining latecomer disadvantages and compete on an equal footing with leaders and followers in high value-added, complex electronics systems and software.

It would be wrong to overemphasize the difficulties facing the latecomers in the future. Some of these appear minor when compared with those already overcome since the 1960s. From a starting position of poverty and backward-

ness the four countries have become fast-growing, highly respected international competitors. Latecomer companies have built up impressive technological competencies, finely tuned to the needs of the most demanding export
markets. As they approach the innovation frontier, the most progressive latecomers have formed long-term strategic alliances with Japanese and US
leaders to jointly develop new technologies. Hundreds of latecomer firms
have narrowed the gap and are poised to make further advances towards the
technology frontier. As this book has shown the achievements of the latecomers are built upon solid historical foundations of learning. While it is impossible to predict the future, it is likely that equally remarkable advances will be
made as new dimensions of latecomer innovation unfold.

NOTES

1. The issue of culture is very controversial and outside the competence of this writer. Some
    argue that a spirit of industrial neo-Confucianism pervades East Asian economic success
    (Vogel 1991). Others point out that Confucianism was often used as an explanation for the
    failure of these same economies in earlier periods and therefore cannot be used as an
    explanation of their success (Riedel 1988 p. 26). See Yahuda (1993) for a critical view of
    cultural explanations.
2. Other types of industries such as petrochemicals, steel and synthetic fibres also have
    different characteristics. South Korea is relatively well documented, especially in the
    heavy and chemical industries. Enos and Park (1988) examine petrochemicals, synthetic
    fibres, machinery, iron and steel. Amsden (1989) looks at automobiles, cement, shipbuild-
    ing, textiles, steel and heavy machinery. Kim and Lee (1987) compare patterns of techno-
    logical change in shipbuilding, cement, automobiles, steel and other sectors.
3. Leading firms such as IBM and Sony are at the frontier. The most advanced latecomers,
    such as Samsung, are at the frontier in some areas (e.g. DRAMs) but not most areas in
    which they compete. At the frontier, firms compete by converting market signals into new
    product designs, requiring extensive, often advanced experimental R&D (and often brand
    leadership for product acceptance). Followers may also compete at the frontier by using
    R&D to generate new designs and to bring novel innovations rapidly to the marketplace.
    Latecomers may carry out R&D, but, by definition, R&D would not yet be central to their
    competitive advantage (e.g. under OEM the key requirement for competitiveness is engi-
    neering capabilities for manufacturing, product–process interfacing, precision engineer-
    ing, product modification and so on). As shown, latecomer R&D tends to be targeted at
    technology catch-up and assimilation rather than moving the innovation frontier forward.
4. The diffusion curve corresponds to the movement of firms through time in relation to
    product and process innovations. The model applies to the electronics industry as a whole,
    rather than to single firms or single products. In the model there is no strict linearity, but,
    rather, a general tendency through time for firms to acquire more complex skills and apply
    these during industrial development. There may be feedback loops between early and late
    stages (e.g. in semiconductors and microwave ovens Samsung began R&D at a very early
    stage, which probably benefited its core manufacturing activities). In addition, as the book
    has shown, not all firms pass through each stage. Once the technology infrastructure is
    sufficiently advanced then new entrants (e.g. ACER of Taiwan) may enter at more ad-
    vanced stages.
5. In Hong Kong clothing was still the largest export sector in the early 1990s.
6. In Hong Kong, watches, clocks and other consumer goods were the largest export items in

the early 1990s, reflecting the weak technological base of the Territory compared with the other three dragons.

7. Forrest (1991) provides a useful summary of simple and complex innovation models incorporating feedback loops and external factors such as policy and the macroeconomy. Magaziner and Patinkin (1989) show how R&D began very early on in the case of Samsung in microwave production.

8. See *Business Week* (June 28 1993 pp. 36–8).

9. Most East Asian innovation has centred upon producing goods and systems new to latecomer firms rather than the marketplace, meeting the broader definition of innovation put forward in Chapter 3. Such innovation has transformed latecomer companies and entire industries, producing significant economy-wide industrial progress in each of the four dragons.

10. See, for example, Utterback and Abernathy (1975), Abernathy and Clark (1985) and Utterback and Suarez (1993). Similar views of how product life cycles should occur can also be found in the marketing literature (e.g. Kotler 1976).

11. See Abbeglen and Stalk (1985) for Japanese corporate innovation patterns.

12. This is a very rough approximation and only applies to electronics. Also note that degrees of policy intervention and rates of FDI varied through time and according to individual sectors and changing government policies.

13. During the 1980s the government intervened much less directly, as the *chaebol* grew in size and competence (see Chapter 4).

14. While import protection was followed in South Korea and Taiwan, the practice took place within an overall framework of export-led industrialization. This coupling of export-led growth with import restriction (which also occurred in Japan), contrasts sharply with the Latin American and Indian approaches, where the local market was the primary focus of policy, rather than the export market.

15. This is an important point as many observers look to South Korea for economic policy lessons. However, in consumer electronics, computing, clothing, footware, bicyles and other important export sectors small firms in Taiwan matched or exceeded the performance of the *chaebol* in export value.

16. Although there are significant differences between the two groups, as discussed earlier (Whitley 1992).

17. Yuan and Low (1990) discuss the balance between private and state entrepreneurship and explain the historical choices of the government.

18. Entrepreneurial failure, a particular form of market failure, occurs when firms are insufficient in numbers or dynamism (or both) to lead industrialization.

19. In electronics the small private firms which took the lead received little direct support from government.

20. See for example World Bank (1993).

21. Writing in 1994, Powell and Takayama (1994 pp. 22–9) argued that Japan's unprecedented three year slump, plus the Yen appreciation, was a structural rather than a cyclical crisis, requiring a long-term restructuring of the economy and calling lifetime employment into question. They cite a study by S.G. Warburg Securities in Tokyo which estimated in 1994 that if the currency continued to appreciate most manufacturers would be forced to relocate more production overseas to survive. Whether cyclical or structural, the difficulties proved especially intense for loss-making Japanese manufacturers of electronics.

22. See Schlendler (1993 pp. 18–23) for details.

23. Ibid.

Bibliography

Abegglen, J.C. (1994), *Sea Change: Pacific Asia as the New World Industrial Center*, New York: The Free Press.

Abegglen, J.C. and Stalk, G.S. (1985), *Kaisha, The Japanese Corporation*, New York: Basic Books.

Abernathy, W.J., Clark, K.B. and Kantrow, A.M. (1983), *Industrial Renaissance: Producing a Competitive Future for America*, New York: Basic Books.

Abernathy, W.J. and Clark, K.B. (1985), 'Innovation: Mapping the Winds of Creative Destruction', *Research Policy*, **14**, pp. 3–22.

Abernathy, W.J. and Utterback, J.M. (1979), 'Patterns of Industrial Innovation', *Technology Review*, **80**, No. 7 (June–July), pp. 41–7.

Akamatsu, K. (1956), 'A Wild Geese Flying Pattern of Japanese Industrial Development: Machine and Tool Industries', *Hitotsubashi Review*, **6**, No. 5, pp. 55–87 (in Japanese).

Amsden, A. (1989), *Asia's Next Giant: South Korea and Late Industrialization*, New York: Oxford University Press.

Ansoff, H.I. and Stewart, J.M. (1967), 'Strategies for a Technology-Based Business', *Harvard Business Review*, **45**, No. 6, pp. 71–83.

Antonelli, C. (1991), *The Diffusion of Advanced Telecommunications in Developing Countries*, Paris: OECD.

Appelbaum, R.P. and Henderson, J. (eds) (1992), *States and Development in the Asia Pacific Rim*, London: Sage Publications.

Archambault, E.J. (1991), 'Small is Beautiful, Large is Powerful: Manufacturing Semiconductors in South Korea'. Unpublished MSc. thesis, Science Policy Research Unit, University of Sussex, England.

Archambault, E.J. (1992), 'Incremental or Radical Change in Semiconductor Manufacturing? Some Evidence from South Korea'. Mimeo, Science Policy Research Unit, University of Sussex, England.

Arrow, J.K. (1962), 'The Economic Implications of Learning by Doing', *Review of Economic Studies*, **XXIX**, pp. 155–73.

Asian Business, various issues.

Bello, W. and Rosenfeld, S. (1991), *Dragons in Distress: Asia's Miracle Economies in Crisis*, London: Penguin Books.

Bloom, M. (1989), 'Technological Change and the Electronics Sector: Perspectives and Policy Options for the Republic of Korea'. Report prepared for Development Centre Project, May, 'Technological Change and the Electronics Sector – Perspectives and Policy Options for Newly Industrialising Economies', Paris: OECD.

Bloom, M. (1991), 'Globalisation and the Korean Electronics Industry, Presentation to the EASMA Conference', 'The Global Competitiveness of Asian and European Firms', Fontainbleau, 17–19 October.

Burgleman, R.A. and Rosenbloom, R.S. (1989), 'Research on Technological Innovation', *Management and Policy*, **4**, pp. 1–23, JAI Press Inc.

Business Korea, various issues.

Business Week, various issues.

Byun, B.M. and Ahn, B.H. (1989), 'Growth of the Korean Semiconductor Industry and its Competitive Strategy in the World Market', *Technovation*, **6**, pp. 635–56.

Chang, P., Shih, C. and Hsu, C. (1993), 'Taiwan's Approach to Technological Change: The Case of Integrated Circuit Design', *Technology Analysis and Strategic Management*, **5**, No. 2, pp. 173–7.

Chaponniere. J.R. (1992), 'The Newly Industrialising Economies of Asia: International Investment and Transfer of Technology', *STI Review*, No. 9, April, Paris: OECD.

Chaponniere, J.R. and Fouquin, M. (1989), 'Technological Change and the Electronics Sector – Perspectives and Policy Options for Taiwan'. Report prepared for Development Centre Project, May, Entitled: 'Technological Change and the Electronics Sector – Perspectives and Policy Options for Newly-Industrialising Economies', Paris: OECD.

Chiang, J.T. (1988), 'Technology Strategies in National Context and National Programs in Taiwan', *Technology in Society*, **10**, pp. 185–204.

Chiang, J.T. (1990), 'Management of National Technology Programs in a Newly Industrializing Country – Taiwan', *Technovation*, **10**, No. 8, pp. 531–54.

China Post, various issues.

Chou, T.C. (1992), 'The Experience of SMEs' Development in Taiwan: High Export-Contribution and Export-Intensity', *Rivista Internazionale di Scienze Economiche e Commerciali*, **39**, No. 12, pp. 1067–84.

Chowdhury, A. and Islam, I. (1993), *The Newly Industrialising Economies of East Asia*, London: Routledge.

Chung, K.H. (1989), 'An Overview of Korean Management', in K.H. Chung and H.C. Lee (eds), *Korean Managerial Dynamics*, New York: Praeger.

Chung, K.H. and Lee, H.C. (1989), 'National Differences in Managerial Practices', in K.H. Chung and H.C. Lee (eds), *Korean Managerial Dynamics*, New York: Praeger.

Clifford, M. (1992a), 'Spring in Their Step', *Far Eastern Economic Review*, 5 November, pp. 57–9.

Clifford, M. (1992b), 'A New Frontier: US Electronics Firm Expands in Asia', *Far Eastern Economic Review*, 17 September, pp. 62–4.

Clifford, M. (1993), 'Growing Pains: Indonesian Growth Zone Makes Patchy Progress', *Far Eastern Economic Review,* 20 May, pp. 60–63.

Computrade International, various issues.

Cowley, A. (1991), 'A Survey of Asia's Emerging Economies: Where Tigers Breed', *The Economist,* 16 November, pp. 5–24.

Dahlman, C.J., Ross-Larson, B. and Westphal, L.E. (1985), *Managing Technological Development: Lessons from the Newly Industrializing Countries*, Washington, DC: The World Bank.

Dahlman, C.J. and Sananikone, O. (1990), 'Technology Strategy in the Economy of Taiwan: Exploiting Foreign Linkages and Investing in Local Capability'. Preliminary Draft, Washington, DC: The World Bank.

Davies, J. (1988), 'The Singapore Vision: An Information-Based Economy', *Journal of Information Science Principles and Practice,* (Netherlands) Vol. **14**, No. 4, pp. 237–42.

Dixon, C. and Drakakis-Smith, D. (eds) (1993), *Economic and Social Development in Pacific Asia*, Routledge: London.

Dodgson, M. (1991), 'Technological Collaboration and Organisational Learning'. DRC Discussion Paper, Science Policy Research Unit, University of Sussex, England.

Dore, R. (1987), *Taking Japan Seriously: A Confucian Perspective on Leading Economic Issues*, London: The Athlone Press.

Dorfman, N.S. (1987), *Innovation and Market Structure: Lessons from the Computer and Semiconductor Industries*, Cambridge, Mass.: Ballinger.

Dosi, G. (1982), 'Technological Paradigms and Technological Trajectories: A Suggested Interpretation of the Determinants and Directions of Technical Change', *Research Policy*, Vol. **11**, No. 3, pp. 147–63.

Dunning, J.H. (1975), 'Explaining Changing Patterns of International Production: In Defence of the Eclectic Theory', *Oxford Bulletin of Economics and Statistics*, Vol. **41**, pp. 269–95.

The Economist, various issues.

EDB (1992a), *Economic Development of Singapore, 1960 to 1991*, Singapore: Economic Development Board.

EDB (1992b), *The Electronics Industry in Singapore,* Singapore: Economic Development Board.

EDB (1992c), *The Disk Drive Industry in Singapore, Special Report*, Singapore: Economic Development Board.

Egan, M.L. and Mody, A. (1992), 'Buyer-Seller Links in Export Development', *World Development*, Vol. **20**, No.3, pp. 321–34.

EIU (1990), *South Korea, No.1, 1990*, Economist Intelligence Unit, Country Report.

Electronic Business, various issues.

Electronic Business Asia, various issues.

Electronics, various issues.

Electronics Times, various issues.

Enos, J.L. and Park, W.H. (1988), *The Adoption and Diffusion of Imported Technology: The Case of Korea*, London: Croom Helm.

Ernst, D. and O'Connor, D. (1992), *Competing in the Electronics Industry: The Experience of Newly Industrialising Economies*, Development Centre, Paris: OECD.

Executive Yuan (1988), 'Science and Technology Advisory Group of the Executive Yuan', Taipei, Taiwan.

Executive Yuan (1989), 'Science and Technology Advisory Group of the Executive Yuan', Taipei, Taiwan.

Far Eastern Economic Review, various issues.

Fok, J.T.Y. (1991), 'Electronics', *Doing Business in Today's Hong Kong, 4th Edition*, American Chambers of Commerce, December.

Forbes, 'Bury Thy Teacher', December 1992, p. 90.

Forrest, J.E. (1991), 'Models of the Process of Technological Innovation', *Technology Analysis and Strategic Management*, 3, No. 4, pp. 439–453.

Fortune (1991), 'Special Report: The Growing Power of Asia', October 7, pp. 118–166.

Fortune (1993), 22 March.

Fransman, M. and King, K. (eds) (1984), *Technological Capability in the Third World*, London: Macmillan.

Freeman, C. (1974), *The Economics of Industrial Innovation*, Penguin Modern Economics Texts, Middlesex: Penguin Books.

Freeman, C. (1987), *Technology Policy and Economic Performance: Lessons from Japan*, London: Frances Pinter Publishers.

Freeman, C. (1990), 'Networks of Innovators: A Synthesis of Research Issues'. Paper presented at the International Workshop on Networks of Innovators organized by the Center for Research and Technology (CREDIT), UQAM, Concordia,University of Montreal, Canada, May 2–3.

Fukasaku, K. (1991), 'Economic Regionalization and Intra-Industry Trade: Pacific Asian Perspectives'. First Draft, Benchmark Study, Research Programme on Globalisation and Regionalisation, Development Centre, Paris: OECD.

Galenson, W. (1992), *Labor and Economic Growth in Five Asian Countries*, New York: Praeger.

Gee, S. (1989), *The Status and an Evaluation of the Electronics Industry in Taiwan*. Report prepared for Development Centre Project, 'Technological Change and the Electronics Sector – Perspectives and Policy Options for Newly Industrialising Economies', Paris: OECD.

Gee, S. (1991), 'Taiwan Enterprises Challenges and Responses Under World Economic Globalisation and Regionalisation', Benchmark Study, Research Programme on Globalisation and Regionalisation, Development Centre, Paris: OECD.

Gerschenkron, A. (1962), *Economic Backwardness in Historical Perspective*, Cambridge, Mass.: Harvard University Press.

Gerstenfeld, A. and Wortzel, L.H. (1977), 'Strategies for Innovation in Developing Countries', *Sloan Management Review*, Fall, pp. 57–68.

Gilbert, A.L. (1990), 'Information Technology Transfer: The Singapore Strategy', in M. Chatterji (ed.), *Technology Transfer in the Developing Countries*, London: Macmillan.

Gregory, G. (1985), *Japanese Electronics Technology: Enterprise and Innovation*, Chichester: John Wiley and Sons.

Gwynne, P. (1993), 'Directing Technology in Asia's Dragons', *Research and Technology Management*, Vol. **36**, No.2, March–April, pp. 12–15.

Haggard, S. (1990), *Pathways from the Periphery: The Politics of Growth in the Newly Industrializing Countries*, London: Cornell University Press.

Henderson, J. (1989), *The Globalisation of High Technology Production: Society, Space and Semiconductors in the Restructuring of the Modern World*, London: Routledge.

Hirschman, A.O. (1958), *The Strategy of Economic Development*, New Haven: Yale University Press.

HKPC (1982), *Study on the Hong Kong Electronics Industry*, Hong Kong Productivity Centre, Hong Kong: Industry Development Board.

Hobday, M.G. (1990), *Telecommunications in Developing Countries: The Challenge from Brazil*, London: Routledge.

Hobday, M.G. (1991), *Country Report on the Republic of Korea*, European Community MONITOR-SAST Report, 'The Needs and Possibilities for Cooperation Between Selected Advanced Developing Countries and the Community in the Field of Science and Technology', SAST Project No. 1, Commission of the European Communities, EUR14141 EN.

Hobday, M.G. (1993), 'The Industrial Technology Research Institute, Taiwan', in E. Arnold, J. Bessant, M. Hobday, R. Murray and H. Rush, *Background/Benchmark Study for the Venezuelan Institute of Engineering*, Centre for Business Research, University of Brighton, Sussex.

Holden, T. and Nakarmi, L. (1992), 'How Japan is Keeping the Tigers in a Cage', *International Business Week*, 11 May, pp. 24–5.

Hone, A. (1974), 'Multinational Corporations and Multinational Buying Groups: Their Impact on the Growth of Asia's Exports of Manufactures – Myths and Realities', *World Development*, **2**, No. 2, pp. 145–9.

Hughes, H. (1988), *Achieving Industrialisation in East Asia*, Cambridge: Cambridge University Press.

IEEE (1991), 'Special Report: AsiaPower', *IEEE Spectrum*, June, pp. 24–66.

III (1988), *Information Industry Yearbook*, published by the Institute for Information Industry, Taiwan.

III (1991), *Information Industry Yearbook*, published by the Institute for Information Industry, Taiwan.

International Management, October 1984.

James, W. (1990), *Basic Directions and Areas for Cooperation: Structural Issues of the Asia-Pacific Economies*, Asia Pacific Cooperation Forum, Session 2, June 21–2, Seoul: Korea Institute for International Economic Policy.

Jansson, H. (1994), *Transnational Corporations in Southeast Asia: An Institutional Approach to Industrial Organisation*, Edward Elgar, London.

Johnstone, B. (1989), 'Taiwan Holds its Lead, Local Makers Move into New Systems', *Far Eastern Economic Review*, 31 August, pp. 50–51.

Jun, Y.W. and Kim, S.G. (1990), 'The Korean Electronics Industry – Current Status, Perspectives and Policy Options'. Report prepared for Development Centre Project, entitled: 'Technological Change and the Electronics Sector – Perspectives and Policy Options for Newly-Industrialising Economies (NIEs), Paris: OECD.

Kamien, M.I. and Schwartz, N.L. (1982), *Market Structure and Innovation*, Cambridge: Cambridge University Press.

Keesing, D. (1988), *The Four Successful Exceptions – Official Export Promotion and Support for Export Marketing in Korea, Hong Kong, Singapore and Taiwan, China*, UNDP–World Bank Trade Expansion Programme, Occasional Paper 2, Washington, DC.

Kelly, B. and London, M. (1989), *The Four Little Dragons: A Journey to the Source of the Business Boom Along the Pacific Rim*, New York: Simon and Schuster Inc.

Kim, C.O., Kim, Y.K. and Yoon, C.B. (1992), 'Korean Telecommunications Development: Achievements and Cautionary Lessons', *World Development*, Vol. **20**, No. 12, pp. 1829–941.

Kim, D.K. and Kim, L. (eds) (1989), *Management Behind Industrialisation: Readings in Korean Business*, South Korea: Korean University Press.

Kim, L. (1989), 'Korea's National System for Industrial Innovation'. Paper prepared for the National Technical System Conference at the University of Limburg in Maastricht, the Netherlands, November 3–4.

Kim, L. and Dahlman, C.J. (1992), 'Technology Policy for Industrialisation: An Integrative Framework and Korea's Experience', *Research Policy*, **21**, pp. 437–52.

Kim, L. and Lee, H. (1987), 'Patterns of Technological Change in a Rapidly Developing Country: A Synthesis', *Technovation*, **6**, pp. 261–76.

Kim, Y. W. (1994), 'Industrialization and Human Resource Development:

Korea's Experience', Science and Technology Policy Institute, Seoul, ROK. Paper prepared for the ASCA Seminar on Science and Technology and Regional Innovation, March 17–18.

K-JIST (1994), *Kwang-Ju Institute of Science and Technology – Status Report*, Seoul: K-JIST.

Koh, A.T. (1987), 'Saving, Investment and Entrepreneurship', in L.B. Krause, A.T. Koh and T.Y. Lee, *The Singapore Economy Reconsidered*, Singapore: Institute for Southeast Asian Studies.

Koh, D.J. (1992), 'Beyond Technological Dependency, Towards an Agile Giant: The Strategic Concerns of Korea's Samsung Electronics Co. for the 1990s'. Unpublished MSc. thesis, Science Policy Research Unit, University of Sussex, England.

Korea Business World, various issues.

Korea Development Bank (1988), *Industry in Korea*, Seoul.

Korea Economic Weekly, 15 March 1993.

Korea Exchange Bank (1980), *The Korean Economy*, Seoul.

Korea Times, various issues.

Kotler, S. (1976), *Marketing Management: Analysis, Planning and Control*, third edn., London: Prentice-Hall International.

Kraar, L. (1993), 'How Samsung Grows so Fast', *Fortune*, 3 May, pp. 26–30.

Krause, L.B. (1987), 'Thinking About Singapore', in L. B. Krause, K.A. Tee and L.T. Yuan, *The Singapore Economy Reconsidered*, Singapore: Institute for South East Asian Studies.

Kwan, C.H. (1994), *Economic Interdependence in the Asia-Pacific Region: Towards a Yen Bloc*, London: Routledge.

Lall, S. (1982), *Developing Countries as Exporters of Technology*, London: Macmillan.

Lall, S. (1992), 'Technological Capabilities and Industrialization', *World Development*, **20**, No. 2, pp. 165–86.

Langlois, R.N., Pugel, T.A., Hacklisch, C.S., Nelson, R.N. and Egelhoff W. G. (1988), *Microelectronics: An Industry in Transition*, Boston, Unwin Hyman.

Lee, H.C. (1989), 'Managerial Characteristics of Korean Firms', in K.H. Chung and H.C. Lee (eds), *Korean Managerial Dynamics*, New York: Praeger.

Levy, B. (1988), 'Korean and Taiwanese Firms as International Competitors: The Challenges Ahead', *Columbia Journal of World Business*, Spring, pp. 43–51.

Lim, Y. (1992), 'Export-Led Industrialization: the Key Policy for Successful Development?', Global Issues and Policy Analysis Branch, UNIDO. Paper prepared for Wilton Park Conference, 14–18 December, London, UK.

Lundvall, B. (1988), 'Innovation as an Interactive Process: From User–

Producer Interaction to the National System of Innovation', in G. Dosi, C. Freeman, R. Nelson, G. Silverberg and L. Soete (eds), *Technical Change and Economic Theory*, London: Frances Pinter.

Mackie, J.A.C. (1992), 'Overseas Chinese Entrepreneurship', *Asian-Pacific Economic Literature*, **6**, No. 1, pp. 41–64.

Magaziner, I.C. and Patinkin, M. (1989), 'Fast Heat: How Korea Won the Microwave War', *Harvard Business Review*, January–February, pp. 83–92.

Malerba, F. (1992), 'Learning by Firms and Incremental Technical Change', *The Economic Journal*, **102**, July, pp. 845–59.

Marshall, A. (1890), *Principles of Economics*, London: Macmillan.

Metcalfe, J.S. (1981), 'Impulse and Diffusion in the Study of Technical Change', *Futures*, **13**, No. 5, pp. 347–59.

Miller, R., Hobday, M.G., Leroux-Demers, T. and Olleros, X., (1995), 'Innovation in Complex Systems Industries: The Case of Flight Simulation', *Industrial and Corporate Change*, **4**, No. 2 forthcoming.

MOST (1993), *Science and Technology in Korea*, Seoul: Ministry of Science and Technology.

Motorola Briefing, January 1993.

Myers, S. and Marquis, D.G. (1969), *Successful Industrial Innovations: A Study of Factors Underlying Innovation in Selected Firms*, National Science Foundation, NSF 69–17.

National Computer Board (1992), *A Vision of an Intelligent Island*, Singapore: National Computer Board.

Nelson, R.R. (ed.) (1993), *National Innovation Systems: A Comparative Analysis*, New York: Oxford University Press.

Nelson, R.R. and Rosenberg, N. (1993), 'Technical Innovations and National Systems', in R.R. Nelson (ed.), *National Innovation Systems: a Comparative Analysis*, New York: Oxford University Press.

NSTB (1991), *Science and Technology: Windows of Opportunity*, Singapore: National Technology Plan, National Science and Technology Board.

O'Connor, D. and Wang, C. (1992), 'European and Taiwanese Electronics Industries and Cooperation Opportunities'. Paper presented at Sino-European Conference on Economic Development, May.

Odagiri, H. and Goto, A. (1993), 'The Japanese System of Innovation: Past, Present, and Future', in R.R. Nelson (ed.), *National Innovation Systems: A Comparative Analysis*, New York: Oxford University Press.

Okimoto, D.I. (1989), *Between MITI and the Market: Japanese Industrial Policy for High Technology*, Stanford, Cal.: Stanford University Press.

Okimoto, D.I. and Rohlen, T.P. (eds.) (1988), *Inside the Japanese System: Readings on Contemporary Society and Political Economy*, Stanford, Cal.: Stanford University Press.

Onn, F.C. (1991), *Impact of New Information Technology in the Banking and*

214   Bibliography

Insurance Industries in Malaysia and Singapore, World Employment Pro-
gramme, Working Paper, Geneva: International Labour Organisation.
Paek, U.C. (1992), *Technology Transfer and Patterns of Technological Inno-
vation: The Case of South Korea*, Seoul: Korea Academy of Industrial
Technology.
Paisley, E. (1993), 'Innovate, Not Imitate', *Far Eastern Economic Review,* 13
May, pp. 64–70.
Pavitt, K. (1984), 'Chips and Trajectories: How Does the Semiconductor
Influence the Sources and Directions of Technical Change?' Mimeo, Sci-
ence Policy Research Unit, University of Sussex, England.
Pavitt, K. (1991), 'What Makes Basic Research Economically Useful?' *Re-
search Policy*, **20**, pp. 109–19.
Pavitt, K. and Rothwell, R. (1976), 'A Comment on "A Dynamic Model of
Process and Product Innovation"', *OMEGA, The International Journal of
Management Science*, **4**, No. 4, pp. 375–7.
Perez, C. (1985), 'Microelectronics, Long Waves and World Structural Change:
New Perspectives for Developing Countries', *World Development*, **13**, No.
3, pp. 441–63.
Perez, C. and Soete, L. (1988), 'Catching up in Technology: Entry Barriers
and Windows of Opportunity', in G. Dosi, C. Freeman, R. Nelson, G.
Silverberg and L. Soete (eds), *Technical Change and Economic Theory*,
London: Frances Pinter.
Porter M.E. (1985), *Competitive Advantage: Creating and Sustaining Supe-
rior Performance*, New York: The Free Press.
Porter M.E. (1990), *The Competitive Advantage of Nations*, London: Macmillan.
Powell, B and Takayama, H. (1994), 'Unbelievable', *Newsweek*, 11 July,
pp. 22–9.
Rana, P.B. (1990), 'Shifting Comparative Advantage Among Asian and
Pacific Countries', *The International Trade Journal*, **4**, No. 3, Spring,
pp. 243–58.
Redding, S.G. (1991), *The Spirit of Chinese Capitalism*, New York: Walter de
Greuter.
Rhee, Y.W., Ross-Larson, B. and Pursell, G. (1984), *Korea's Competitive Edge:
Managing the Entry into World Markets,* Baltimore: John Hopkins Press.
Riedel, J. (1988), 'Economic Development in East Asia: Doing What Comes
Naturally?', H. Hughes (ed.), *Achieving Industrialization in East Asia*,
Cambridge: Cambridge University Press.
Rush, H., Hobday, M., Bessant, J. and Arnold, E. (1994), 'The Benchmarking
of National Science and Technology Institutes – Strategies for Best Prac-
tice'. Mimeo, University of Brighton/Science Policy Research Unit, Uni-
versity of Sussex, England.

Sakong, I. (1993), *Korea in the World Economy*, Washington, DC: Institute for International Economics.

Schive, C. (1990), *The Foreign Factor: The Multinational Corporation's Contribution to the Economic Modernization of the Republic of China*, Stanford, Cal.: Hoover Institution Press.

Schlendler, B.R. (1993), 'Japan: Hard Times for High Technology', *Fortune*, 22 March, pp. 18–23.

Schmookler, J. (1966), *Invention and Economic Growth*, Cambridge, Mass.: Harvard University Press.

Schumpeter, J.A. (1950), *Capitalism, Socialism, and Democracy*, third edn, New York: Harper and Row.

Singapore Electronics Manufacturers Directory, various issues.

Singapore Investment News (1991), *Precision Engineering Supplement*, Singapore: Economic Development Board.

Singapore Investment News (1992), February, Economic Development Board, Singapore.

Singapore Times, 7 October 1991.

Sisodia, R.S. (1992), 'Singapore Invests in the Nation-Corporation', *Harvard Business Review*, May–June, pp. 40–50.

Soete, L. (1985), 'International Diffusion of Technology, Industrial Development and Technological Leapfrogging', *World Development*, Special Issue on Microelectronics, **13**, No. 3, pp. 409–23.

Soon, T.T. and Huat, T.C. (1990), 'Role of Transnational Corporations in the Transfer of Technology to Singapore', in M. Chatterji (ed.), *Technology Transfer in the Developing Countries*, London: Macmillan.

SPRU (1972), *Success and Failure in Industrial Innovation, Report on the Project SAPPHO by the Science Policy Research Unit*, University of Sussex, London: Centre for the Study of Industrial Innovation

Suh, S.C. (1974), 'The Korean Electronics as Export Industry'. Paper presented at the Korea Development Institute-HIID Conference, June 25–8, Seoul, Korea.

Suh, S.C. (1975), 'Development of a New Industry Through Exports: The Electronics Industry in Korea', in W. Hong and A.O. Krueger (eds), Trade and Development in Korea. Proceedings of a Conference held by the Korea Development Institute, 1975, Seoul, Korea.

Sunday Morning Post (1993), *China Business Review*, Hong Kong, 30 May 1993.

Swinbanks, D. (1993), 'What Road Ahead for Korean Science and Technology?', *Nature*, **364**, July, pp. 377–84.

Sylla, R. and Toniolo, G. (1991) 'Introduction: Patterns of European Industrialization During the Nineteenth Century', in R. Sylla and G. Toniolo (eds), *Patterns of European Industrialization*, London: Routledge.

TEAMA (1991), *Taiwan Electric and Electronic Products Buyers Guide, 1991 to 1992*, Taipei: Taiwan Electric Appliance Manufacturers' Association.

TEAMA Directory 1992, Taipei: Taiwan Electric Appliance Manufacturers' Association.

Teece, D. J. (1986), 'Profiting From Technological Innovation: Implications for Integration, Collaboration, Licensing and Public Policy', *Research Policy*, **15**, pp. 285–305.

Telecom Sources (1993), January, Hong Kong.

Tho, T.V. and Urata, S. (1991), 'Emerging Technology Transfer Patterns in the Pacific Asia'. Paper presented at International Conference Entitled: 'The Emerging Technological Trajectory of the Pacific Rim', held at Fletcher School of Law and Diplomacy, Tufts University, October 4–6.

Thoburn, J.T., Leung, H. M., Chau, E. and Tang, S. H. (1991), 'Investment in China by Hong Kong Companies', *IDS Bulletin,* **22**, No 2, Institute of Development Studies, Sussex, England.

Tushman, M. and Anderson, P. (1986), 'Technological Discontinuities and Organizational Environments', *Administrative Science Quarterly*, **31**, pp. 439–65.

United Nations (1988) *Handbook of International Trade and Development Statistics*, UN.

Urata, S. (1990), *The Rapid Globalization of Japanese Firms in the 1980s: An Analysis of the Activities of Japanese Firms in Asia*, Research Programme on Globalisation and Regionalisation, Development Centre, Paris: OECD.

Utterback, J.M. and Abernathy, W.J. (1975), 'A Dynamic Model of Process and Product Innovation', *OMEGA, The International Journal of Management Science*, **3**, No. 6, pp. 639–56.

Utterback, J.M. and Suarez, F.F. (1993), 'Innovation: Competition, and Industry Structure', *Research Policy*, **15**, pp. 285–305.

Vatikiotis, M. (1993), 'Chip off the Old Block: Doubts Plague Singapore-Centred "Growth Triangle"', *Far Eastern Economic Review,* 7 January, p. 54.

Vernon, R. (1960), *Metropolis 1985: An Interpretation of the Findings of the New York Metropolitan Region Study*, Cambridge, Mass.: Harvard University Press.

Vernon, R. (1966), 'International Investment and International Trade in the Product Life Cycle', *Quarterly Journal of Economics*, **80**, No. 2, pp. 190–207.

Vernon, R. (1975), 'The Product Life Cycle Hypothesis in a New International Environment', *Oxford Bulletin of Economics and Statistics*, **41**, pp. 255–67.

Vogel, E.F. (1991), *The Four Little Dragons: The Spread of Industrialization in East Asia*, Cambridge Mass.: Harvard University Press.

Wade, R. (1990), *Governing the Market: Economic Theory and the Role of Government in East Asian Industrialization*, Princeton, New Jersey: Princeton University Press.

Wellenius, B., Miller, A. and Dahlman, C.J. (eds) (1993), *Developing the Electronics Industry: A World Bank Symposium*, Washington, DC: World Bank.

Westphal, L.E., Kim, L. and Dahlman, C.J. (1985), 'Reflections on the Republic of Korea's Acquisition of Technological Capability', in N. Rosenberg and C. Frischtak (eds), *International Transfer of Technology: Concepts, Measures, and Comparisons*, New York: Praeger.

Whitley, R. (1992), *Business Systems in East Asia: Firms, Societies and Markets*, London: Sage Publications.

World Bank (1993), *The East Asian Miracle: Economic Growth and Public Policy*, World Bank, New York: Oxford University Press.

Wortzel, L. H. and Wortzel, H.V. (1981), 'Export Marketing Strategies for NIC and LDC-based Firms', *Columbia Journal of World Business*, Spring, pp. 51–60.

Yahuda, M. (1993), 'Perspective: East Asia Special', *The Times Higher Education Supplement*, 26 November, p. 17.

Yamashita, S. (1991), 'Japan's Role as a Regional Technological Integrator in the Pacific Rim'. Paper presented at the conference on 'The Emerging Technological Trajectory of the Pacific Rim', Tufts University, Medford, Mass., October 4–6.

Yearbook of World Electronic Data (1988 and 1990) Vols 1 and 2, Benn Electronics/Elsevier.

Yuan, L.T. and Low, L. (1990), *Local Entrepreneurship in Singapore: Private and State*, Singapore: The Institute of Policy Studies, Times Academy Press.

Yue, C.S. (1985), 'The Role of Foreign Trade and Investment in the Development of Singapore', in Galenson, W. (ed.), *Foreign Trade and Investment: Economic Development in the Newly Industrializing Asian Countries*, Wisconsin: University of Wisconsin Press.

Index